Guidance on Using Corridor and Subarea Planning to Inform NEPA

Federal Highway Administration

April 5, 2011

Executive Summary

This guidance is provided to assist transportation planners and environmental practitioners in the use of corridor and subarea planning to inform the National Environmental Policy Act (NEPA) review process.[1] Current law provides authority for, and even encourages, the integration of information and products developed in highway and transit planning into the NEPA review process.[2] This document responds to the need for additional guidance on how best to use corridor and subarea planning to bridge the transportation planning and NEPA processes as described in Appendix A to 23 Code of Federal Regulations (CFR) Part 450 – Linking the Transportation Planning and NEPA Processes.[3]

States, metropolitan planning organizations (MPOs), and local governments have primary responsibility for transportation planning. The transportation planning process required by 23 U.S.C. §§ 134 and 135 and 49 U.S.C. §§ 5303-5306 sets the stage for future development of transportation projects.[4] Federally-funded highway and transit projects originate in the statewide and metropolitan transportation planning processes.[5] A State, MPO, or public transportation operator may undertake a multimodal, systems-level corridor or subarea planning study as part of the statewide and metropolitan transportation planning process. The results or decisions of this study may be used as part of the overall project development process consistent with NEPA and FHWA regulations. Often, since it happens later in the project development process, the environmental analysis done to meet NEPA requirements for transportation projects is largely disconnected from the planning process. This may result in planning decisions being overlooked or disregarded under NEPA. When decisions are revisited, it can lead to misapprehension, duplication of work, added expense, or confusion for stakeholders.

Corridor and subarea plans are conceptual level planning studies, which focus on a particular corridor or region and can help determine where there is a transportation need. The transportation regulations governing the use of corridor and subarea studies identify products from this type of planning that may be used to inform NEPA, including, the purpose and need or goals and objectives statement(s); the general travel corridor and/or general mode(s) definitions; the preliminary screening of alternatives and elimination of unreasonable alternatives; the basic description of the environmental setting; and/or the preliminary identification of environmental impacts and environmental mitigation.[6] The regulations lay out the conditions that must be met in order for these planning products to be used in Federal Highway Administration (FHWA) and Federal Transit Administration (FTA) NEPA evaluations. The most important of these are agency and public involvement and good documentation.

When conducting corridor and subarea planning, it is important to have involvement from a broad range of partners, including resource and regulatory agencies, NEPA practitioners, planning and development partners, legal counsel, and the public. Proper documentation that explains the thought process behind planning decisions is also essential. Corridor and subarea studies are not

---

[1] 43 U.S.C. § 4321 et. seq.
[2] *See* Environment and Planning Linkage Processes Legal Guidance. http://www.fhwa.dot.gov/hep/plannepalegal050222.htm.
[3] Appendix A to 23 CFR Part 450. http://edocket.access.gpo.gov/2007/07-493.htm.
[4] See also implementing regulations at 23 CFR Part 450, http://edocket.access.gpo.gov/2007/07-493.htm.
[5] 23 CFR Part 450. http://edocket.access.gpo.gov/2007/07-493.htm.
[6] 23 CFR §§ 450.212 and 450.318. http://edocket.access.gpo.gov/2007/07-493.htm.

the only approach to link planning and NEPA, but they provide many benefits. Corridor and subarea studies can help agencies identify efficiencies, enhance flexibility, build understanding between agencies, and respond to fiscal challenges. There is no guarantee that what is decided in corridor and subarea planning will be advanced into project development, but using corridor and subarea studies to inform the NEPA process provides the opportunity to identify issues of concern early and build project understanding among agency stakeholders and the public.

TABLE OF CONTENTS

1.0 INTRODUCTION 1
   1.1 Need for Guidance 1
      1.1.1 Transportation Planner Perspectives 2
      1.1.2 NEPA Practitioner Perspectives 2
   1.2 Process for Developing this Guidance 3
   1.3 Using the Guidance 3
   1.4 Connecting Planning Studies with NEPA 4
   1.5 Using Corridor and Subarea Planning to Inform NEPA 4

2.0 PLANNING AND INITIATING A STUDY 7
   2.1 What is corridor and subarea planning? 7
      2.1.1 Typical Elements of a Corridor Study 7
      2.1.2 Subarea Planning 8
   2.2 When to perform a planning study? 9
   2.3 How to fund a planning study? 9
   2.4 What do you hope to accomplish in a planning study? 10
      2.4.1 Advantages and Disadvantages 10
      2.4.2 Address Fiscal Challenges 11
      2.4.3 Integration with Other Planning 12
   2.5 Who should be involved? 12
      2.5.1 Resource and Regulatory Agencies 13
      2.5.2 Transportation NEPA Practitioners 13
      2.5.3 Planning and Development Partners 14
      2.5.4 Legal Counsel 14
      2.5.5 Other Stakeholders 15

3.0 CONDUCTING A STUDY 16
   3.1 Things to Consider 16
   3.2 Products 17
      3.2.1 Purpose and Need or Goals and Objectives Statements 17
      3.2.2 General Travel Corridor and/or General Modes Definition 18
      3.2.3 Preliminary Screening of Alternatives and Elimination of Unreasonable Alternatives 19
      3.2.4 Basic Description of the Environmental Setting 19
      3.2.5 Preliminary Identification of Environmental Impacts and Environmental Mitigation 20
   3.3 Making the Connection and Carrying Information through the Process ...21
      3.3.1 Early Considerations 21
      3.3.2 Using the Notice of Intent to Link Planning and NEPA 22

4.0 MAKING A PLANNING STUDY VIABLE FOR NEPA 24
    4.1 Weight Given to Planning Products Informing NEPA 24
    4.2 Appropriate Environmental Analysis 24
    4.3 Good Documentation 25
        4.3.1 Planning/ Environmental Linkages Questionnaire 25
        4.3.2 Corridor Planning Study Checklist 29
        4.3.3 Issue/Concern Tracking and Response 32
    4.4 NEPA Practitioner's Perspective 32

5.0 LESSONS LEARNED 34

6.0 USING TIERING TO CONNECT PLANNING WITH NEPA 35

7.0 CONCLUSION 36

APPENDIX A: LEGAL, POLICY AND GUIDANCE FRAMEWORK 37
    A.1 NEPA and Transportation Decisionmaking 37
    A.2 FHWA/FTA Regulatory Language (as of the date of this guidance) 38
    A.3 Other Laws, Regulations, and Orders Governing Environmental Decisionmaking ... 41
    A.4 Current Policy Framework 43
    A.5 NCHRP Report 435: Guidebook for Transportation Corridor Studies 44

APPENDIX B: CASE STUDIES 46
    B.1 Libby North Corridor Study 46
    B.2 Parker Road Corridor Study 47
    B.3 Regional Outer Loop Corridor Feasibility Study 48

APPENDIX C: RESOURCES 49
    C.1 Corridor and Subarea Planning 49
    C.2 Federal Regulations and Guidance 49
    C.3 Planning and Environment Linkages 50
    C.4 State Project Examples 51
    C.5 Related Research 53

List of Acronyms and Abbreviations

AASHTO	American Association of State Highway Transportation Officials
CDOT	Colorado Department of Transportation
CE	Categorical Exclusion
CEQ	Council on Environmental Quality
CFR	Code of Federal Regulations
CRP	Community Resource Panel
DEIS	Draft Environmental Impact Statement
DOT	Department of Transportation
DVRPC	Delaware Valley Regional Planning Commission
EA	Environmental Assessment
EIS	Environmental Impact Statement
EPA	Environmental Protection Agency
ESA	Endangered Species Act
FHWA	Federal Highway Administration
FTA	Federal Transit Administration
FWS	Fish and Wildlife Service
GIS	Geographic Information System
HUD	Housing and Urban Development
ICC	Intercounty Connector
IPWG	Integrated Planning Work Group
ITD	Idaho Transportation Department
MDT	Montana Department of Transportation
MOA	Memorandum of Agreement
MOU	Memorandum of Understanding
MPO	Metropolitan Planning Organization
NCHRP	National Cooperative Highway Research Program
NCTCOG	North Central Texas Council of Governments
NEPA	National Environmental Policy Act
NHS	National Highway System
NOI	Notice of Intent
PEL	Planning and Environment Linkages
PennDOT	Pennsylvania Department of Transportation
PL	Metropolitan Planning
REAP	Regional Ecological Assessment Protocol
SAFETEA-LU	Safe, Accountable, Flexible, Efficient Transportation Equity Act: A Legacy for Users
SEIS	Supplemental Environmental Impact Statement
SPR	Statewide Planning and Research
STIP	Statewide Transportation Improvement Program
STP	Surface Transportation Program
TCS	Tiering, Corridor, and Subarea Studies
TIP	Transportation Improvement Program
U.S.C.	United States Code
UDOT	Utah Department of Transportation
WFRC	Wasatch Front Regional Council
WSDOT	Washington State Department of Transportation

1.0 INTRODUCTION

1.1 NEED FOR GUIDANCE

States, MPOs, and local governments have primary responsibility for transportation planning. The transportation planning process required by 23 U.S.C. §§ 134 and 135 and 49 U.S.C. §§ 5303-5306 sets the stage for future development of transportation projects and subsequent project-level decisions. Despite the importance of the transportation planning process, the environmental analyses done to meet NEPA requirements have often been conducted largely disconnected from the public and stakeholder agency input and transportation analyses used to develop long-range plans, statewide/ metropolitan Transportation Improvement Programs (STIPs/TIPs), and/or planning-level corridor and subarea studies.

The NEPA project development process is intended to inform the public and help Federal officials make decisions based on an understanding of the environmental consequences of a proposed action, and take actions that protect, restore, and enhance our environment.[7] NEPA analysis typically adds more specificity and technical information to State and local planning-level analyses, but the goal is the Federal NEPA review will not unnecessarily revisit these analyses and decisions. Revisiting of planning analyses and decisions often results from a need to develop or document information during NEPA that should more appropriately have been developed and documented during planning. When this occurs, it results in a duplication of work, more expense, confusion for the public and policymakers, and a potential delay in project implementation (see Figure 1 below).[8]

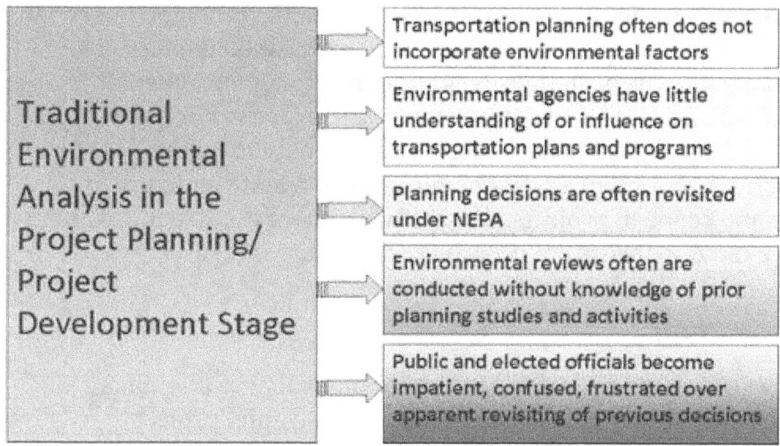

Figure 1: Traditional Environmental Analysis

In February 2007, the Federal Highway Administration (FHWA) and Federal Transit Administration (FTA) issued statewide and metropolitan transportation planning regulations that implemented changes to Federal law as a result of Public Law 105-178, the Transportation Equity Act for the 21st Century (TEA-21) and Public Law 109-59, the Safe, Accountable, Flexible, Efficient Transportation Equity Act: A Legacy for Users (SAFETEA-LU). The transportation planning regulations supplement authority under the Council on Environmental Quality (CEQ) NEPA regulations and allow the FHWA and FTA, as NEPA lead agencies, to use the results or decisions of in State department of transportation (DOT), metropolitan planning organization (MPO), or public

---

[7] CEQ regulation 40 CFR Parts 1500-1508. http://ceq.hss.doe.gov/nepa/regs/ceq/1500.htm#1500.1
[8] Linking the Transportation Planning and National Environmental Policy Act (NEPA) Processes. FHWA/FTA. February 2005.

transportation operator corridor and subarea planning studies as part of the environmental review process under NEPA so long as legal requirements are met.[9] This guidance discusses implementation of the planning requirements and explains the linkage between the transportation planning and project development/NEPA processes as contained in 23 Code of Federal Regulations (CFR) Part 450, including Appendix A.[10]

The statewide and metropolitan transportation planning regulations and Appendix A to 23 CFR Part 450 allow for analysis from corridor and subarea studies to be fully utilized during project environmental review, when conditions in that regulation are satisfied. FHWA has recognized the need for guidance that encourages the use of corridor and subarea studies to provide continuity between the transportation planning and NEPA process. This guidance is meant to meet that need by describing for the transportation planning and environmental communities how corridor and subarea planning studies can best be used to inform the NEPA process – facilitating seamless, coordinated decisionmaking.

1.1.1 Transportation Planner Perspectives

The motivation for initiating corridor and subarea planning varies. For transportation planners at State DOTs and MPOs, it can make sense to conduct a corridor study when there are limited resources (e.g., funding, staff) and a planning-level study can help resolve initial issues. A corridor study can better define the purpose and need for a transportation improvement, or help prioritize among various competing projects or corridors. Subarea studies are often used to help address broader issues of land use, growth management, and resource protection, often through scenario planning, before project-level transportation solutions are identified. Transportation planners can benefit by getting early feedback from resource agencies and environmental stakeholders. This early consultation with State and local planning, economic development, and environmental protection agencies is encouraged under the planning regulations.[11] The expectation is that early consultation will help agencies identify key environmental factors and resources that will lead to more informed decisionmaking. Corridor and subarea studies can also help State and local planners understand the magnitude and scope of projects, and allow planners to learn more about a particular corridor or subarea before moving forward with project development.

1.1.2 NEPA Practitioner Perspectives

NEPA practitioners are increasingly recognizing the value of corridor and subarea planning. Corridor and subarea planning can lead to improved transportation planning and project development. Early environmental analysis and documentation can maximize avoidance of impacts. Early planning can sometimes identify major projects that under closer analysis have more cost-effective solutions with fewer impacts, by transforming a large project into a series of networked improvements to existing facilities. NEPA practitioners are seeing that resource agencies may become involved in corridor and subarea planning if they have available staff to participate and are confident that there is managerial support for this type of collaboration. Under

---

[9] *See* Federal Highway Administration regulations at 23 CFR §§ 450.212, 450.318, and 771.111(a)(2); *see also* Federal Transit Administration 49 CFR Part 613.

[10] The requirements in 23 CFR §§ 450.212 and 450.318 apply only to products of the transportation planning process conducted directly pursuant to 23 U.S.C. §§ 134 and 135 and implementing regulations in 23 CFR Part 450. Those requirements do not apply to other types of materials and the planning regulations do not otherwise affect FHWA's authority to adopt or incorporate materials pursuant to CEQ regulation at 40 CFR Parts 1500-1508.

[11] 23 CFR §§ 450.212(b) and 450.318(b).

the traditional process, project-level decisions may be derailed suddenly by previously unforeseen significant environmental impacts, a lack of political will, or evolving transportation needs that could have been discovered sooner through a corridor or subarea planning process. NEPA practitioners may feel that conducting project planning before initiating project development can be more efficient, lead to avoiding significant impacts, and improve the connection to broader, system-level transportation goals and analysis.

1.2 Process for Developing this Guidance

FHWA developed this guidance based on its experience and in concert with its agency partners. Information resources included the planning and NEPA regulations, FHWA's Planning and Environment Linkages (PEL) initiative, and National Cooperative Highway Research Program (NCHRP) Report 435.[12]

In 2007, FHWA created the Tiering, Corridor, and Subarea Studies (TCS) Group. The TCS Group is a subgroup to the Integrated Planning Work Group (IPWG) established under Executive Order 13274: Environmental Stewardship and Transportation Infrastructure Project Reviews. The Executive Order established an Interagency Task Force that consisted of representatives from the U.S. Departments of Transportation, Interior, and Agriculture, and Environmental Protection Agency (EPA). The Task Force charged the IPWG with identifying the challenges and opportunities inherent to integrated planning – the linkage that occurs when transportation agencies and environmental resource agencies effectively coordinate their planning processes.

Over the past few years, the IPWG has explored how transportation agencies consider environmental concerns early in the planning process and partner with resource agencies to identify strategies to maximize environmental protection and transportation benefits. The TCS Group was asked to focus on the use of planning studies to inform NEPA and develop guidance to better link the transportation planning and NEPA environmental review processes.

In 2009, the group convened a peer exchange in Denver, Colorado that examined the use of corridor planning studies as a foundation for NEPA decisionmaking.[13] The peer exchange highlighted several different approaches that regions have taken in the use of corridor studies. Peers shared lessons they had learned and made recommendations on how best to use corridor planning to bridge the transportation planning and environmental review processes. Input from the peer exchange, the Linking Planning and NEPA Advisory Panel comprised of planning, environmental, and legal staff from FHWA Headquarters, Resource Center, and Division Offices, resource and regulatory agencies, and a multi-agency group of transportation experts organized by FHWA, informed development of the guidance.

1.3 Using the Guidance

This guidance is intended for transportation planners and NEPA practitioners. The goal of this guidance is to both inform and provide practical advice to the transportation and environmental community for developing and utilizing corridor and subarea planning studies during the transportation planning process and incorporating the results into NEPA during project

---

[12] NCHRP 435, "Guidebook for Transportation Corridor Studies: A Process for Effective Decisionmaking," served as a foundational reference for this guidance.
[13] FHWA Peer Exchange on Using Corridor Planning to Inform NEPA, http://www.environment.fhwa.dot.gov/integ/peer_exch_corridors.asp.

development. Project examples, additional resources, and links to further information are provided throughout this guidance and in the attached appendices.

1.4 Connecting Planning Studies with NEPA

There are different approaches transportation agencies and their partners may take in connecting planning studies with NEPA, which generally follow a sequential timeline (see Figure 2 below).

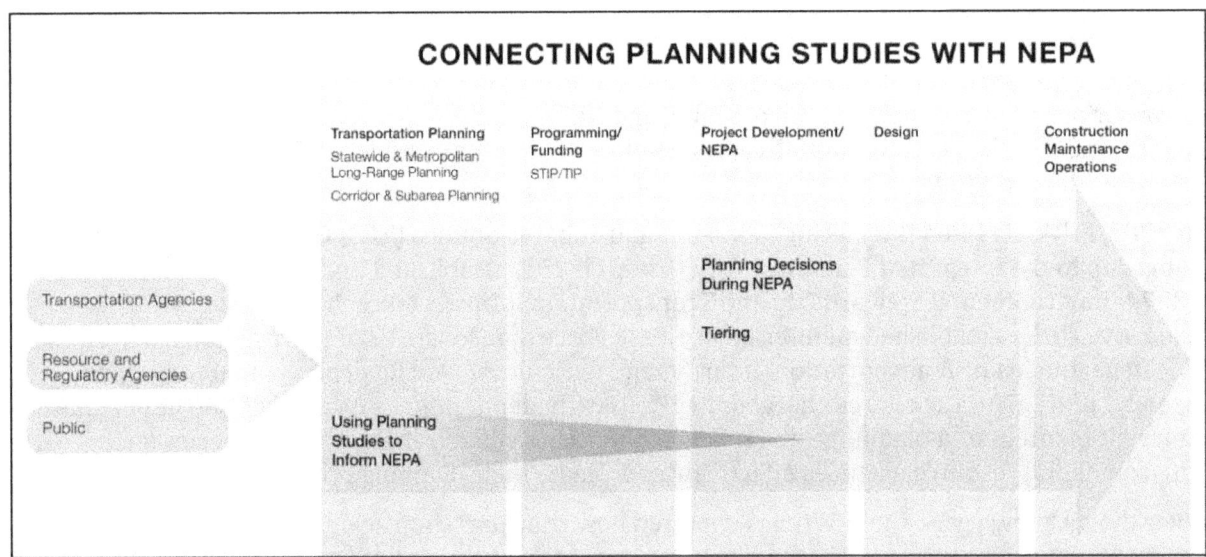

Figure 2: Timing of different approaches available to connect transportation planning with NEPA

Agencies may choose the traditional approach of defining purpose and need and potential alternatives solely from information developed during the formal project development/ NEPA stage. They may opt for a tiered approach under NEPA, or agencies may use corridor and subarea studies developed during planning to inform the NEPA process. Regardless of the approach chosen, an agency may ultimately end up in the same place at the design and construction stage – but there can be varying benefits associated with each approach.

This guidance focuses on the benefits and considerations to use corridor and subarea planning to inform NEPA. Comparisons are made throughout the guidance with the traditional approach, and additional information on tiering is in Chapter 6.

1.5 Using Corridor and Subarea Planning to Inform NEPA

The focus of this guidance is conducting corridor and subarea planning during the transportation planning process and utilizing that information to inform the agencies and the public during the NEPA process. Source material produced by, or in support of, the transportation planning process may be incorporated directly or by reference into subsequent NEPA documents in accordance with the FHWA and CEQ regulations.[14] Under FHWA planning regulations, the NEPA lead agencies must agree that the material's incorporation will aid in establishing or evaluating the purpose and need for the project, reasonable alternatives, cumulative or other impacts on the human and natural environment, or mitigation.[15]

---

[14] 40 CFR §§ 1502.21 and 1502.24; 23 CFR §§ 450.212(b) and 450.318(b).
[15] 23 CFR §§ 450.212(b)(1) and 450.318(b)(1).

Pursuant to FHWA planning regulations[16], to be used in NEPA the systems-level, corridor, or subarea planning study must be conducted with involvement of interested State, local, Tribal, and Federal agencies.[17] This means that those entities must be offered the opportunity to participate in the study and it is recommended that efforts be made to foster strong and continuous participation. There must be public review of the study[18], meaning that the public must be advised of the preparation of the study and where it is available. Both the public and agencies must have a reasonable opportunity to comment during the transportation planning process and development of the corridor or subarea planning study.[19] Documentation of relevant decisions is needed in a form that is identifiable and available for review during the NEPA scoping process.[20] It must be in a format that can be appended to or referenced in the NEPA document.[21] FHWA must review the study.[22] The review is both to inform FHWA of the direction and progress of the study and to give FHWA an opportunity to provide feedback to the study preparers on whether the study work is on track to meet the requirements in the planning regulation.

One disadvantage is that participation by partner resource and regulatory agencies in the planning level study may be limited since there is no requirement for them to participate. Moreover, most resource agencies have limited staff available for participation in transportation planning. If the public process is well-planned and efficient, however, agency and public participation can be strong. Both corridor and subarea planning can be highly effective ways to involve non-traditional transportation partners in early discussions about more sustainable integration of transportation projects with housing, community development, infrastructure, and economic development.

Corridor and subarea studies can be used to produce a wide range of analyses or decisions for FHWA review, consideration and possible adoption in the NEPA process for an individual transportation project, including:[23]

- The foundation for purpose and need statements;
- Definition of general travel corridor and/or general mode(s);
- Preliminary screening of alternatives and elimination of unreasonable alternatives;
- Planning-level evaluation of indirect and cumulative effects;
- Regional or eco-system-level mitigation options and priorities; and
- Linkage with housing, development, economic, and environmental goals and analysis.

It is important to emphasize that analyses done during the transportation planning process do not serve as NEPA compliance. The lead agency (or agencies) in NEPA must choose whether to adopt planning information during the scoping process in order for that information to be used in and comply with NEPA. The products of the transportation planning process – especially if thoroughly documented – can inform an environmental assessment (EA) or environmental impact statement

---

[16] 23 CFR §§ 450.212(b)(2) and 450.318(b)(2).
[17] 23 CFR §§ 450.212(b)(2)(i) and 450.318(b)(2)(i).
[18] 23 CFR §§ 450.212(b)(2)(ii) and 450.318(b)(2)(ii).
[19] 23 CFR §§ 450.212(b)(2)(iii) and 450.318(b)(2)(iii).
[20] 23 CFR §§ 450.212(b)(2)(iv) and 450.318(b)(2)(iv).
[21] Id.
[22] 23 CFR §§ 450.212(b)(2)(v) and 450.318(b)(2)(v).
[23] 23 CFR §§ 450.212(a) and 450.318(a).

(EIS) to meet NEPA requirements. The end result is that the effort in NEPA may be greatly enhanced if project sponsors can rely on previous planning work.

2.0 PLANNING AND INITIATING A STUDY

2.1 WHAT IS CORRIDOR AND SUBAREA PLANNING?

Corridor and subarea planning represent a range of activities and study elements that are useful in bringing information together to help communities make transportation decisions. These planning activities are often an intermediary step between the broader long-range planning process and the more detailed work of project development. Transportation planners often undertake corridor and subarea studies as part of the statewide, metropolitan and/or local transportation planning processes.[24]

Corridor and subarea planning studies are often conceptual level studies that can help determine whether there is a need for a transportation project. The basis for the study is an accurate and inclusive identification of the environmental and community goals for the area. The studies can help to identify the purpose and need or the vision, goals, and objectives of the corridor or subarea. The geographic limits of the study, the basic description of the environmental setting, development trends, or changes in land use, modes, or alternatives may also be identified. A study engages the community and stakeholders in a process of thinking about the area's future and then documents those results as the basis for future planning and project development.

The studies may be used to identify projects for inclusion in the statewide or metropolitan long-range transportation plan. Among the solutions that are often considered are potential improvements on existing facilities. This exploration of solutions, if documented appropriately, may help FHWA to determine whether a categorical exclusion (CE), EA, or EIS will be necessary for a proposed improvement that requires FHWA approval. As a result, corridor and subarea studies are often an attractive way to explore an area's needs and potential solutions using planning funds in preparation for initiating the NEPA process for a proposed project.

> **Corridor Planning**
> The Delaware Valley Regional Planning Commission (DVRPC) completes several corridor planning studies each year based on its congestion management process and long-range plan. Areas of focus for these studies have varied throughout the Philadelphia (PA) metropolitan area.
> http://www.dvrpc.org/corridors/

2.1.1 Typical Elements of a Corridor Study

A corridor study is a targeted analytical study that addresses specific needs of a corridor or particular geographic area. Corridor studies are used to achieve various goals. The content of a corridor study will vary based on the actual corridor itself and the study's purpose, but generally, a corridor study would include:

- A reason for conducting the study, including the main issues affecting system performance;
- A clear definition and justification for the study area boundaries, including a description of corridor resources and potentially affected stakeholders;
- A budget, schedule, and list of expected products arising from the study. Products that may come out of the study include:

---

[24] NCHRP Report 435: Guidebook for Transportation Corridor Studies – A Process for Effective Decision-Making. Transportation Research Board, 1999.

- goals, objectives, and evaluation measures for the corridor;
- alternative strategies to address identified problems;
- an analysis of forecasted impacts of these alternative strategies in terms of environmental, transportation, and financial impacts; and
- an evaluation of how each alternative strategy addresses the specified problems of, and goals and objectives for, the corridor.

Together, these components are used to define a concept and scope for a transportation improvement or set of improvements, including the mode(s), facilities, and general location of the proposed improvement.

Corridor studies done as part of planning can help inform the various elements of the transportation decisionmaking process, such as defining the transportation deficiency or elements of the purpose and need, determining funding needs, or determining how corridor improvements fit into a larger system plan. Multimodal corridor planning on existing roadways can have a broader focus, incorporating improvements to management and operations, transit, bicycle/pedestrian movement, access management, and development of a connected network of streets for local travel. It can also incorporate planning for land use, mixed-use development and transit oriented development, parking management, and other strategies to improve mobility and accessibility while reducing environmental impacts.

Given the various goals a corridor study can achieve, it is important to determine the study's objectives at the outset and then structure the study to reach the desired objectives. For example, the goal may be to more clearly define the project purpose and need, or to simply identify a general travel corridor as part of the long-range planning process.

> **Rosencrans Corridor Mobility Study**
> The City of San Diego's (CA) Rosencrans Corridor Mobility Study examined a four-mile corridor using both extensive stakeholder outreach and objective measurement. The study included developing measures of effectiveness for each mode under each study alternative, which allowed for objective evaluation of proposed alternatives for the corridor. Examining four segments of the corridor and all modes allowed for the identification of specified mobility issues in each area – ranging from relocation of transit stops and traffic calming measures to the need for increased traffic flow in other portions.
> http://www.dot.ca.gov/dist11/departments/planning/pdfs/systplan/RosecransCorridorStudyNoAppendicesCBTPGrantFebruary2010.pdf

2.1.2 Subarea Planning

Subarea planning addresses the development of a defined portion of a region (such as a county) in more detail than area-wide or regional plans. Subarea studies are similar to corridor studies, with the distinction that a subarea study generally addresses more of the total planning context and the broader transportation network for the area. In particular, congestion, land use and housing, growth management, and resource protection, and their interactions with the transportation network, are often part of a subarea study. Subarea studies may incorporate greater participation on the part of potentially affected stakeholders in the study area, in order to build consensus around a more comprehensive vision for the area's future. Subarea studies are often developed by State DOTs and MPOs working together with neighboring localities.

Planning tools like scenario modeling and visualization techniques are often used in subarea studies. These public participation tools help local communities and other stakeholders better understand the interactions between different planning issues in an area (e.g., multimodal transportation connections, housing affordability, energy use, climate change) and the range of possible outcomes for future development. Scenario planning is especially helpful as a decision support tool to help compare the benefits and impacts of different development patterns and transportation investments.

2.2 WHEN TO PERFORM A PLANNING STUDY?

For projects or needs that have been identified in the long-range transportation plan, a corridor or subarea study can be used to better refine the project or need. The results can then feed back into the long-range plan where smaller, more affordable projects identified can be programmed into the TIP. A planning study can also be useful to help define problems or identify potential solutions to carry forward into the NEPA and project development process. A study can assist when funding is limited and decisions are needed as to what improvements can be made in a timely and cost-effective manner. A study is advised if the project is complex: for example, if the project is regionally significant, has environmental constraints, incorporates analysis of housing and community development options, is costly or controversial, or has the potential for many alternatives that could be indistinct and confusing.

A planning study may not be necessary for activities with little to no controversy or no significant impacts. For such projects, it might save time and money to begin NEPA directly. However, projects such as reconstruction, bridge replacement, or a widening should be reviewed on a case by case basis; a planning study could be needed depending on the context of the project and the sensitivity of the location.

2.3 HOW TO FUND A PLANNING STUDY?

A typical corridor or subarea study undertaken by an MPO or State DOT may take advantage of a variety of possible funding sources. The following funding sources are available for planning studies:[25]

Metropolitan Planning (PL): FHWA planning funds designated for MPOs under 23 U.S.C. § 104(f) are available to MPOs in order to carry out the metropolitan transportation planning process required by 23 U.S.C. § 134, including development of metropolitan area transportation plans and TIPs. Eligible activities include conducting inventories of existing routes to determine their physical condition and capacity, determining the types and volumes of vehicles using these routes, predicting the level and location of future population, employment, and economic growth, and using such information to determine current and future transportation needs.

Statewide Planning and Research (SPR): FHWA planning funds designated for States under 23 U.S.C. § 505(a) are available to States in order to carry out the statewide planning process required by 23 U.S.C § 135. Eligible activities include engineering and economic surveys and investigations, the planning of future highway programs and local public transportation systems and the planning of the financing of such programs and systems, including metropolitan and statewide planning under § 134 and § 135 [of 23 U.S.C.], and studies of the economy, safety, and convenience of surface transportation systems and the desirable regulation and equitable taxation of such systems.

Some confusion has been expressed over just how far a PEL study is able to proceed using SPR or PL funding. Because both of these programs are tied to planning and not project development, the

---

[25] Where there are multiple purposes associated with a study or planning activity, so that the work may benefit non-transportation or non-planning activities, consult with the FHWA Division financial manager about whether cost allocation under 2 CFR Part 225 is required.

point where planning ends and project preliminary engineering begins is a critical discussion that needs to occur. Any general inventory data, system-wide level data collection or analysis and how they would be applied to the corridor (no specific alternative selected) would be considered planning. Publishing the Notice of Intent (NOI) and beginning NEPA would be considered a project-level activity and appropriate for project funds. For more detailed discussion on the topic of eligible uses of PL or SPR funding (or any other Federal Aid Fund Source) please speak with your FHWA Division office and consult the current Guide to Federal-Aid Programs and Projects.

Although the boundary between planning and project is a funding eligibility issue for PL/SPR funds, using other Federal-aid program funds that provide inherent funding flexibility should be explored.

Surface Transportation Program (STP): While typically thought of as "project" funds this most flexible FHWA funding source may also be used for surface transportation planning in accordance with 23 U.S.C. § 133(b)(7).

National Highway System (NHS): While typically thought of as "project" funds, this funding program may be used for transportation planning activities associated with the NHS in accordance with 23 U.S.C. § 103(b)(6)(E).

2.4 What do you hope to accomplish in a planning study?

2.4.1 Advantages and Disadvantages

Identify Efficiencies: When corridor and subarea studies are prepared with the environmental review process in mind, upfront planning can lead to efficiencies later in project development and better projects. The transportation improvement can be better designed from the start to address community and environmental needs, and if planned with stakeholder involvement can lead to greater community and agency buy-in for the resulting project. These efficiencies can lead to time and cost savings for developing and evaluating different improvement options. For projects that have long lead times between planning and development, transportation agencies can use corridor and subarea studies to identify needs early on and then conduct more detailed studies and analysis during NEPA. Similarly, for large projects that may not have sufficient funding to be built all at once, a corridor study can set the framework for future detailed studies and eventual improvements, ensuring consistency in the overall objectives and design of the corridor.

Enhance Flexibility: While the scope of a project is being developed, objectives can be explored and defined to meet community and environmental needs and other identified issues. Engaging resource agencies early, before they have entered NEPA "review mode," can lead to projects that better respond to identified environmental issues in their basic scope and design.

Build Understanding: Corridor and subarea studies set the broader context and provide a general understanding among the public and stakeholders about the transportation needs, potential issues, and potential solutions. The studies can create project buy-in and reduce questions during later environmental review. Relationships developed during corridor and subarea planning can form the basis for improved relationships among agencies and encourage coordination among environmental and transportation planners.

No guarantees: Resource agency input during a corridor or subarea study does not guarantee approval of a resulting project, or that all potential issues will be identified during the planning study. Taking a collaborative approach only provides the opportunity for key issues to be identified early and project expectations to be mutually agreed upon.

Coordinate Resources: Resource agencies may be reluctant to get involved in planning studies because of limited staff time and a need to focus their attention on environmental review. One way to address this is through the use of funded positions, where transportation funding pays the cost of resource agency staff working on transportation projects (23 USC §139(j)).

Communicate with the public: Corridor and subarea studies may be confused with project development or near-term improvements. Clearly communicating the planning and project development timeline to the public as part of the corridor or subarea planning process can help avoid confusion.

2.4.2 Address Fiscal Challenges

When faced with fiscal challenges, transportation agencies need to efficiently and effectively prioritize investments. Sometimes agencies initiate NEPA analysis on transportation projects before enough is known about the transportation need and options for addressing it, or about major constraints that will affect the scope and nature of any proposed solution. In such cases, the NEPA process is used to address broad planning-type questions. Corridor and subarea studies can better address these questions by enabling agencies to cost-effectively identify transportation and environmental needs early in planning.

By approaching a project at a more conceptual level, planning studies allow agencies to explore creative and cost-effective solutions that are not strictly transportation-based, such as changing growth patterns to help redirect future demand on the transportation system, to evaluate whether they are reasonable alternatives that merit analysis. Fiscal constraints can also be addressed by breaking larger projects into smaller components that can be completed as funding is available, as long as each component has logical termini and independent utility. Corridor and subarea studies can support projects with budget constraints by establishing a framework for development of the full corridor and the prioritization of component improvements. A planning study may also help agencies find more creative solutions to address an area or corridor's needs. Resulting projects may have a more clearly defined purpose and need, be less expensive, require less environmental review because environmental impacts have been avoided, and will likely have more community support if developed in a participatory manner.

Overcoming Fiscal Challenges

Interstate 83 Master Plan
The Pennsylvania Department of Transportation (PennDOT) originally anticipated it would have to do one large EIS for the I-83 corridor, an 11-mile section of Interstate 83 in Harrisburg, Pennsylvania. A closer look revealed that different locations within the corridor had different needs. PennDOT realized that the cost of the programmed projects for the corridor far exceeded available funding. PennDOT decided to divide the corridor into four independent sections that could be advanced through environmental review and programming independently. The I-83 Master Plan is essentially a framework for all partners to use in the future. As funding becomes available, prime sections of I-83 may be improved.

The Libby North Corridor Study
The Montana Department of Transportation (MDT) had a project in place to reconstruct a 14-mile section of Highway 567 to current standards. This segment abuts a National Forest and has serious endangered species concerns. MDT realized that under traditional project development methods, these environmental concerns would likely require an EIS that would be beyond the project's budget. In response, MDT stepped back and launched a corridor study with full community involvement. The community and resource agencies indicated that rather than a full rebuild, they wanted to maintain the rural character of the road with safety improvements and minimal environmental impact. The resulting project avoided significant environmental impacts and was able to shift from a costly EIS to a CE. For more information, see Appendix B, Case Studies.

2.4.3 Integration with Other Planning

There is increasing recognition of the need to link transportation planning with other planning done to ensure that communities develop sustainably and efficiently. While transportation agencies only have jurisdiction over the transportation network, transportation problems are often the result of local and regional land use decisions and conventional development patterns, which shape how and when people use the transportation system. The environmental impact of transportation and land use decisions also needs to be better addressed at the planning stage, by integrating the planning efforts of transportation, land use, housing, and environmental agencies. Planning studies may include existing transportation facilities, land use, general socio-economic information, other transportation facilities (including ports, airports, and rail lines), environmental, and geological features. Analysis should be conducted to the level of detail needed to define the problem. Information sharing through integrated planning can lead to more informed decision-making, reduced duplication of effort, and better understanding of goals across agencies. Variables that can impact the level of analysis in the study may include size of community, demographics and other socio-economic factors, distribution of population and major employers, growth trends and projections, physical condition of the transportation infrastructure, traffic and safety, environmental setting and topography, and environmental justice populations.

At the Federal level, the HUD-DOT-EPA Interagency Partnership for Sustainable Communities is demonstrating and promoting a more integrated approach to planning and development. States and regions are also being encouraged to adopt this approach. Corridor and subarea studies offer an opportunity for planners to not only inform NEPA but also other related efforts, such as land use, housing, transit, and sustainable communities. Better linking land use and housing development with transportation investments can have measurable environmental benefits.

Many States and MPOs have already undertaken corridor planning studies that successfully integrate transportation with land use and other planning efforts. These studies address the transportation needs of a corridor along with the needs of the communities within that corridor. Smaller projects that can be implemented faster are developed by regional partnerships in conjunction with local land use planning and zoning authorities. By integrating planning efforts both across sectors and across planning levels (local to MPO corridor level), transportation and land use development inform and support one another.

> **Places29**
> The Thomas Jefferson Planning District Commission (TJPDC), the MPO for the Charlottesville (VA) area, partnered with Virginia DOT and neighboring jurisdictions in an effort to combine land use and transportation planning. Places29 built off the US29N Corridor Study and established a vision for accommodating growth along US Route 29, incorporating placemaking and transportation solutions into MPO and Albemarle County master plans. TJPDC used an FHWA Eco-Logical grant to integrate transportation planning with environmental resource management. They gathered existing resource data and assigned weighted values, resulting in a regional Ecological Value Map. This included an example "Least Environmental Cost Alignment" analysis for a new roadway proposed in Places29.
> www.tjpdc.org/transportation/places_29.asp
> http://www.tjpdc.org/environment/index.asp

2.5 WHO SHOULD BE INVOLVED?

It is essential to start the planning process with all partners at the table – the public, local governments, transportation agencies, resource agencies, and other stakeholders. As part of the long-range transportation planning process, DOTs and MPOs should consult (as appropriate) with various State and local agencies responsible for land use, natural resources, environmental protection, conservation, and historic preservation. Bringing these same players into the corridor planning process can yield better planning recommendations and help build relationships between agencies that support further integrated planning efforts.

2.5.1 Resource and Regulatory Agencies

Early consultation with resource and regulatory agencies can help integrate resource agency goals and plans into the transportation planning process. Consultation may involve comparisons of transportation plans with State conservation plans and inventories of natural/historical resources. Resource agencies have in-depth knowledge of the environmental issues that may affect an area and may have more up-to-date information available than is contained in their agency plans.

Regional Outer Loop Corridor Feasibility Study

This study is an evaluation for the Dallas-Fort Worth region of the need and feasibility for a 240-mile outer loop using a network of transportation routes to improve regional mobility, freight flows, and economic vitality. The study is in its early stages, but the North Central Texas Council of Governments has already begun collaborating with resource agencies for their expertise, technical tools, and information. Some of their collaboration and use of technical tools includes:

- Integrated Stakeholder Coordination Efforts: early and continuous information exchange and partnership activities to integrate environmental planning factors into all study phases.

- Regional Ecological Assessment Protocol (REAP): a planning and screening level assessment tool that uses geographic information system(GIS) to classify land based on ecological importance;

- GISST: a GIS-driven environmental assessment and data management tool that uses over 100 different types of environmental resource criteria to score and assesses potential environmental impacts; and

- NEPAssist: an innovative Web-based tool that draws environmental data dynamically from EPA regions' GIS databases and provides immediate screening of environmental assessment indicators for a user-defined area. These features contribute to a streamlined review process that potentially raises important environmental issues at the earliest stages of project development.

For more information, see Appendix B, Case Studies.

Resource agencies can not only share their data but their technical expertise. Planning agencies can draw on this expertise when acquiring, understanding, and using environmental data. This has certain advantages. For example, sharing data supports quantitative and defensible planning approaches that identify, coordinate, and analyze existing data with tools such as GIS. A number of useful software tools can help incorporate land use, economics, and ecological/ geophysical modeling into the planning process. The flexible approach and structure of newer tools is suitable for planning, and many can be used by GIS experts as well as non-experts with a minimum of training and support.

2.5.2 Transportation NEPA Practitioners

Transportation NEPA practitioners typically focus on environmental analysis and review, and are not involved in the preparation of transportation planning documents. If NEPA practitioners are invited to become involved in transportation planning studies, and understand the value and intended use of planning information for informing future NEPA review, the result may be a better and more efficient project delivery. During NEPA, transportation planning study results will need to be assessed to determine the suitability of the documentation for use in the NEPA process. It is important to engage NEPA practitioners to ensure that they are aware of the authority that local,

MPO, and State governments have for planning decisions and also are aware of the transportation planning decisionmaking authority that is contained in the statewide and metropolitan transportation planning regulations.

2.5.3 Planning and Development Partners

Partner agencies and local staff involved in land use planning, community development, and housing will often have experience working in the study area as well as knowledge of plans and policies that affect the area. Engaging planning and development partners builds relationships that can be valuable in integrating transportation and land use solutions, and continue when the project is actually built. Collaboration with these agencies can help ensure that the resulting transportation plans are consistent with and supportive of other plans for the area. These partners may be able to supply information needed for transportation models and analyses. Additionally, these agencies and staff may have experience working in the community and may be able to identify important local stakeholders and long-standing issues.

The building and development industry and landowners are also important stakeholders to involve in the corridor or subarea planning process both for their knowledge and their support. They will often have valuable insights about the current and future conditions of the local land and development market, and, once the plan is completed, these are the stakeholders that implement future development in the area.

2.5.4 Legal Counsel

Involvement of legal counsel early in the planning process can be of great value when an agency intends to use planning products to inform NEPA. Counsel can help agencies identify and resolve potential issues, and provide guidance to assist agencies' development of planning documentation that meets requirements for use in later NEPA proceedings. Counsel input can be particularly helpful with respect to planning work intended to help identify purpose and need and a range of alternatives for detailed NEPA evaluation.

As agencies prepare planning products for possible future use in the NEPA environmental review and project development process, there needs to be an awareness of how the documentation, including relevant planning material, could be incorporated into the planning administrative record and, in the event of litigation, potentially into the agency's administrative record which is filed with the court. In general, all planning analyses and decisions that are used in subsequent project-level environmental work need to be well documented and included in the project's administrative record. In the event of litigation challenging a Federal NEPA decision, the administrative record will always include the NEPA documents themselves. Types of planning documents that could become part of the project's administrative record if relied on in the NEPA review include technical reports, such as corridor and subarea studies, meeting summaries that document coordination with resource agencies, the public, and other stakeholders, telephone memos that reference conversations held, correspondence with agencies and stakeholders, and comment/response matrices that track all of the comments received and demonstrate how they have been addressed. Legal counsel can provide other suggestions and can advise best what needs to be done to ensure the project's administrative record is in good form and is sufficient.[26]

---

[26] For more information, see AASHTO Practitioner's Handbook 01, Maintaining a Project File and Preparing an Administrative Record for a NEPA Study. AASHTO Center for Environmental Excellence. July 2006.

2.5.5 Other Stakeholders

As in all good planning work, it is crucial to have early and extensive outreach to stakeholders, including the general public, elected officials, advocacy groups, businesses, and other interested parties. An opportunity for public involvement is necessary if the study is to be used as part of NEPA.[27] It may be useful to develop a Public Involvement Plan for the study. An inclusive public involvement process not only improves the likelihood that the study will be acceptable for use during NEPA, it more importantly leads to better projects that the public supports. Stakeholders are those potentially impacted by a project, not just those within an agency's geographic jurisdiction. Public participation is a continuing need. As metropolitan areas continue to grow and change, the public that should be consulted can change within a region.

Robust outreach and consensus-building as part of a corridor or subarea study ensures that the full range of community issues, opportunities, and ideas are brought to the table while there is still the flexibility to incorporate and address them. A collaborative approach that builds consensus also helps to avoid unexpected challenges to future project development. For controversial projects, the study can build consensus on the framework for measuring project impacts and outcomes that, if well-documented and adhered to, can help agencies strike a balance between competing needs.

Documentation and a good communication strategy can be essential in promoting public involvement and in addressing expectations throughout the planning/project development process. Visual tools and creative methods for outreach can help draw stakeholders to participate and help clarify the process for those who are not familiar with transportation planning language.

Effective Approach to Public Involvement

US-20 Corridor Project

The US-20 Corridor Plan, from Ashton Hill Bridge to the Montana State Line, is a long-range planning effort by the Idaho Transportation Department (ITD) to assess the condition of the US-20 Corridor and identify necessary improvements to meet the corridor's system and user needs for the next 20 years. From the outset, ITD's district office in charge of the plan emphasized having an interactive, ongoing planning process that focused on listening to stakeholders, capturing needs, and explaining the planning process so better decisions could be reached. To address ongoing concerns about endangered species, the district formed an action team of experts with mitigation, stream-banking, and permitting expertise. That team was then able to transition to design issues when funding unexpectedly became available for improvements.
http://itd.idaho.gov/Projects/

---

[27] 23 CFR §§ 450.212(b)(2)(i) and 450.318(b)(2)(i).

3.0 CONDUCTING A STUDY

The transportation planning regulations governing the use of transportation planning materials to inform project development (23 CFR 450.212 and 450.318, *see also* Appendix A.2) identify the following five items among the products that corridor or subarea studies may produce for a proposed transportation project:

- Purpose and need or goals and objectives statement(s);
- General travel corridor and/or general mode(s) definition (e.g., highway, transit, or a highway/transit combination);
- Preliminary screening of alternatives and elimination of unreasonable alternatives;
- Basic description of the environmental setting; and/or
- Preliminary identification of environmental impacts and environmental mitigation.

These products may be incorporated directly or by reference into NEPA documents, provided certain conditions are met.

3.1 THINGS TO CONSIDER

In order for a corridor or subarea planning study to be used in NEPA pursuant to FHWA regulations at 23 CFR §§ 450.212 and 450.318, certain conditions in those regulations must be met. This is because the information and products coming from the planning process must be sufficiently comprehensive and accurate so that the Federal government may reasonably rely upon them in its NEPA analysis, documentation, and decision-making.

Under FHWA's planning regulations, the agencies leading the NEPA process must agree that using the planning material will help inform NEPA as described below.[28] Transportation planning agencies should consider the level of detail that the corridor or subarea planning study should entail. The planning material may include identification or evaluation of such matters as:

- the project's purpose and need,
- reasonable alternatives for the project,
- the project's impacts on the environment, and/or
- how to mitigate the project's impacts.

The statewide and metropolitan planning regulations, 23 CFR §§ 450.212 and 450.318 do not dictate whether or not a previously-developed or related study should be used to inform NEPA. However, the environmental impact regulations, 23 CFR § 771.111, specify that early coordination with the appropriate agencies and the public aids in determining the type of NEPA document required, the scope of the document, the level of needed analysis, and the related environmental requirements.

At the outset of entering NEPA for EIS projects, the lead agencies may include a reference to a prior planning study and products from that study in the project's NOI to prepare an EIS. The NOI may reference the specific study, and it may include the project draft purpose and need and range of alternatives that were developed in the planning study.

---

[28] 23 CFR §§ 450.212(b)(1) and 450.318(b)(1).

Corridor and subarea planning studies are prepared as part of the transportation planning process.[29] Accordingly the study process must be consistent not only with 23 CFR §§ 450.212 and 450.318, but also with general transportation planning requirements such as providing an opportunity for public involvement and considered relevant planning factors.[30] Interested State, local, Tribal and Federal agencies should be included in the transportation planning process, and must be given a reasonable opportunity to comment upon the long-range transportation plan and TIP.[31] Any work from the transportation planning process that is carried into the NEPA process must have been documented and available for public review during the study process.[32] An opportunity for public involvement is required[33] and, ideally, the planning study was conducted with strong participation from other agencies and from the public, State, local, Tribal, and Federal agencies or departments, and particularly environmental, regulatory, and resource agencies with jurisdiction or an interest in the area of study. For those involved, the goal is to have early and meaningful involvement throughout the process. For a systems-level, corridor, or sub-area study to be included in NEPA, review by FHWA and FTA must occur during the corridor or subarea study process.[34] The review may be accomplished by providing a copy of the draft study to FHWA or FTA (as applicable).

> **Auke Bay Corridor Study**
>
> The Auke Bay Corridor Study by Alaska Department of Transportation and Public Facilities (DOT&PF) included an extensive public involvement plan meant to "verify the basis for the project" and "establish the legitimacy of problem solving and decisionmaking processes." Involvement went beyond public meetings to include a citizens' advisory committee, a newsletter and project mailing list, as well as a media strategy. By actively engaging with the media and citizens, Alaska DOT encouraged public understanding and analysis of proposed alternatives.
> http://dot.alaska.gov/stwdplng/projectinfo/ser/abcorr/index.shtml

Finally, the results and decisions of the planning process need to be documented in a way that is clear, suitable, and readily available for incorporation into the NEPA document.[35] If a study or decision is to be used in a NEPA review, the study and the documented decision will need to be publicly available for those wishing to comment on the NEPA document, so it is important to maintain public access to the planning documents until the NEPA process is complete.[36]

3.2 PRODUCTS

3.2.1 Purpose and Need or Goals and Objectives Statements

The purpose and need statement in a NEPA document is where the planning process and the NEPA document most clearly intersect. A sound planning process is a primary source of the project purpose and need. When defining its vision, a community develops long-range goals and forecasts the needs of the system for the future. It prioritizes strategies for addressing those needs and proposes a timeframe in which to develop those strategies into actual projects. Corridor and

---

[29] 23 CFR §§ 450.212(a) and 450.318(a).
[30] See, e.g,, 23 CFR §§ 450.206, 450.208, 450.210, 450.306, and 450.316.
[31] 23 CFR §§ 450.210) and 450.316.
[32] 23 CFR §§ 450.212(b)(2) and 450.318(b)(2).
[33] 23 CFR §§ 450.212(b)(2) (ii) and 450.318(b)(2)(ii).
[34] 23 CFR §§ 450.212(b)(2)(v) and 450.318(b)(-2)(v).
[35] 23 CFR §§ 450.212(b)(2) (iv) and 450.318(b)(2)(iv).
[36] 40 CFR § 1501.21 and 23 CFR §§ 450.212(b)(2) (iv) and 450.318(b)(2)(iv).

subarea planning help a community envision its future transportation system in a more specific, localized way than long-range transportation planning. The resulting project information provides the basis on which to build a purpose and need statement under NEPA.[37] Corridor and subarea studies help refine a project's purpose and need in two main ways:

- Defining the *goals and objectives* or vision statement for a particular area or corridor and,
- Framing the *scope* of the problem to be addressed by a future project.

State and metropolitan management systems or processes for congestion, pavement, bridges, and safety can all produce analyses that help shape the purpose and need statement. Both fiscal constraints and management systems analyses can inform and be informed by corridor studies, so this information may be incorporated into a corridor study, as well as other planning documents.

The purpose and need or goals and objectives statements, like the other planning products arising from corridor and subarea studies, do not have to be at the same level of detail as those provided under NEPA. The purpose and need statement:

- Should be a statement of the transportation problem (not a statement of a solution);
- Should be based on articulated planning factors and developed through a certified planning process;
- Should be specific enough so that the range of alternatives developed will offer real potential for solutions to the transportation problem;
- Must not be so specific as to "reverse engineer" a solution[38]; and
- May reflect other priorities and limitations in the area, such as environmental resources, growth management, land use planning, and economic development.

3.2.2 General Travel Corridor and/or General Modes Definition

The results of a well-supported and documented corridor or subarea study may be used to define the State's or MPO's desired general travel corridor and/or general modes for a future transportation improvement. The general travel corridor is not the specific alignment, but does direct future study of the corridor into one general area. A recommendation of the general mode(s) to be used as the transportation solution focuses on what modes can meet the goals and objectives identified for the area or corridor. For example, a corridor study may conclude that transit or a combination of highways and transit are the only modes that will meet the future needs of that corridor. The planning study does not need to identify both general travel corridor and general mode; it may identify only one of these, or neither.

> **Interstate-405 Corridor Study**
> Washington State Department of Transportation (WSDOT) worked with cities, counties, Federal agencies, transit agencies and community groups to develop consensus for a long-term vision for the multi-modal redevelopment of Interstate-405, the major travel corridor east of Seattle (WA). The I-405 Corridor Study culminated in a series of improvements for the corridor, including identifying bus rapid transit as the appropriate mass transit mode for the corridor.
> http://www.wsdot.wa.gov/projects/i405/

---

[37] For a discussion of the roles and responsibilities of Federal, State, and local agencies in the planning and NEPA contexts, *see Citizens for Smart Growth v. Peters*, 716 F.Supp.2d 1215, 1222-1225 (S.D. Fla. 2010).
[38] *See, e.g., Citizens Against Burlington, Inc. v. Busey*, 938 F.2d 190, 196 (D.C. Cir. 1991).

The level of detail in the planning study will vary based on the context, facility, and resources (time, technology) available to the agency conducting the study. For instance, in selecting a preferred general mode, a planning decision could be made based on typical benefits and standard costs of each mode type. Similarly, general travel corridor selection could be made based on avoidance of known critical habitats or treasured resources, rather than on field inspections of all potential corridors. During the NEPA process, additional analysis may be needed, building from the planning products.

3.2.3 Preliminary Screening of Alternatives and Elimination of Unreasonable Alternatives

There are two ways that a corridor or subarea study can have an effect on the screening of alternatives. First, the planning study can provide information the NEPA lead agencies may decide to use to develop the purpose and need statement, which can then be used to identify preliminary alternatives for analysis (see Section 3.2.1). Second, the corridor or subarea study may be used to directly evaluate alternatives and suggest elimination from detailed NEPA study of alternatives that are not reasonable (as that term is defined in the NEPA context). If this is a part of the corridor study, there should be detailed documentation of the alternatives that the State or MPO wish to eliminate and the reasons for their elimination. The study may then be incorporated by reference into the later NEPA process if it meets requirements previously discussed, including the interagency and public coordination required during NEPA. To do this, the planning process should identify and study alternatives, similar to the approach used in a preliminary screening of alternatives in the NEPA process. The results of the planning analysis may be used later in NEPA to support a Federal decision to eliminate from further study any alternatives that are not feasible or do not meet the project purpose and need. Public involvement and input from environmental, regulatory, and resource agencies is particularly important in the second approach. Documentation of their involvement will strengthen the validity of the planning study. This will help to demonstrate that alternatives were not eliminated without proper consultation and support from interested and affected parties. It is critical to properly document the analysis, public and agency involvement, and resulting planning decisions to ensure that these analyses meet requirements for use in the NEPA process.

When screening alternatives as part of a corridor or sub-area planning study, the level of detail in the analysis will be higher than the level of detail typically used in a planning document. Any resulting transportation planning decisions that have eliminated alternatives should have a rational basis that has been thoroughly documented, including documentation of the necessary and appropriate public involvement processes. Still, it should be made clear in planning that there is no guarantee that an alternative will not resurface during the NEPA process.

3.2.4 Basic Description of the Environmental Setting

Corridor and subarea studies will generally address the context and some of the potential impacts associated with proposed transportation improvements. These can be valuable inputs to the discussion of the affected environment and environmental consequences during NEPA analysis. Planning-level information and analyses that can be useful in that context include:

- Regional development and growth analyses;
- Local land use, growth management, or development plans and projections of future land use, natural resource conservation areas, and development;
- Demographic trends and forecasts, including population and employment projections;

- GIS overlays showing past, current, or predicted future conditions of the natural and built environments;
- Environmental scans that identify environmental resources and environmentally sensitive areas;
- Descriptions of airsheds and watersheds; and
- The outputs of natural resource planning efforts, such as wildlife conservation plans, watershed plans, special management areas, and multiple species habitat conservation plans.

Typically, a transportation planning study documenting the existing environment is not detailed or current enough to meet NEPA standards; it may need to be supplemented during the NEPA process. The planning study should provide enough detail to support the analyses conducted in the study, and as much as possible document the project-level environmental setting. When scenario analysis is used in a corridor or subarea study, the resulting model outputs, coupled with GIS layer mapping, can provide an appropriate level of detail for planning-level discussion and transportation planning decisions. The maps and visualizations used in scenario planning are often useful in supporting informed public engagement in planning decisionmaking.

3.2.5 Preliminary Identification of Environmental Impacts and Environmental Mitigation

While planning studies will generally not determine in detail what the impacts of a future project would be, these studies can be an effective basis for consideration of direct, indirect and cumulative impacts in NEPA analysis. As noted earlier, corridor and subarea studies should provide an overview of the planning area's current and future development patterns, growth, and demographics. By describing the interconnections between the transportation system, community resources, and the environment and natural ecosystem, the planning process provides a baseline for measuring how the current environment will change and helps to identify what those changes may look like.

> **Arkansas' Ecoregion-Based Approach to Wetlands Mitigation**
> To help offset wetland losses, the Arkansas Highway and Transportation Department (AHTD) set up a number of wetland mitigation banks throughout the state. By taking a landscape-scale or eco-region approach and contributing a larger bank site rather than many small sites into the overall ecology of the area, the project provided ecological connectivity and prevented further environmental fragmentation. This is an example of mitigation which could be defined during the planning process.
> http://environment.fhwa.dot.gov/ecosystems/eei/ar05.asp

Planning studies can be used to avoid and/or minimize environmental effects through the use of early screening. They can also be used to start interagency discussions to develop advance mitigation agreements or create mitigation banks. For these purposes, it is especially important to include the use of products from the environmental and natural resource expertise and data. By utilizing the analyses of both environmental data and transportation planning information, planners can screen planning-level decisions, such as the general travel corridor, for their impact on watersheds or habitat areas. Knowing the potential impacts earlier allows agencies to avoid impacts and, for unavoidable impacts, develop more effective and economical mitigation strategies achieve both environmental and transportation objectives.

The information generated during a planning study needs to be detailed enough to support planning-level decisions for environmental impact avoidance, minimization, early and compensatory mitigation. Under 23 CFR §§ 450.214(j) and 450.322(f)(7), long range transportation plans require a discussion of types of potential environmental mitigation activities as part of the statewide long range transportation plan and the metropolitan transportation plan, including those that have the greatest potential to restore and maintain the environmental functions affected by planning. The discussion shall be developed in consultation with Federal, State, and Tribal land management, wildlife, and regulatory agencies.[39]

Transportation planning studies that include consideration of mitigation on a regional or watershed level can help facilitate the development of early mitigation planning which could be implemented prior to traditional transportation project milestones such as compensatory mitigation.[40] Typically, environmental impacts and environmental mitigation will need to be studied and analyzed in more detail during the NEPA analysis.

3.3 MAKING THE CONNECTION AND CARRYING INFORMATION THROUGH THE PROCESS

An essential component in linking planning activities to the NEPA process is making sure that the information developed during the planning stage is carried through to project development. To do this, it is important to properly document the planning study, meet the conditions set out by the regulations for incorporation of planning products,[41] and build relationships between planning agencies, resource agencies, and the stakeholders that will be conducting and reviewing the environmental documentation. This will help ensure that all interested parties are aware of the planning study and are comfortable with using it to inform NEPA.

3.3.1 Early Considerations

For a NEPA practitioner, there are several factors to consider when determining whether or not to use a planning study in NEPA. These factors, along with the joint lead agencies' intention to incorporate the planning study decisions, should be discussed with the relevant agencies during NEPA project scoping. These factors include:

- The age, relevance, and reliability of the planning study, its data, and its analysis;

- Whether assumptions made in the study are consistent with those to be used in the NEPA analysis;

- Inclusion of relevant stakeholders in the planning process, and how well the links and distinctions between the planning and NEPA processes were explained;

- Availability of the planning document for review and/or incorporation into the NEPA document; and

---

[39] 23 CFR §§ 450.214(j) and 450.322(f)(7).
[40] Early mitigation planning often occurs in connection with potential impacts regulated under the Clean Water Act (33 U.S.C. § 1344) and/or the Rivers and Harbors Act (33 U.S.C. §§ 401, 403) and implementing regulations at 40 CFR Part 230 and 33 CFR Parts 325 and 332.
[41] 23 CFR §§ 450.212 and 450.318, 40 CFR § 1502.21, and 23 CFR § 771.111(a)(2).

- How well transportation, natural resource and land use plans inform one another as a part of the general transportation planning process.

When deciding whether to use a planning-level screening of alternatives, there are additional considerations for a NEPA practitioner. When the planning study is incorporated by reference, the NEPA document will need to:

- Identify the alternatives eliminated during the planning process, including the broad categories of alternatives eliminated by a study's definition of a general travel corridor or general modes;
- Summarize the reasons for the elimination of those alternatives; and
- Summarize the analysis and document the FHWA evaluation that supports the elimination of alternatives by referencing relevant sections of the planning study and then accurately incorporating the study into the NEPA document by reference or by appending it.

In the case where alternatives are screened or eliminated, it is important to note that any reasonable alternatives that remain after the planning study will need to be studied as part of the NEPA process, even if they are not the preferred alternative identified in the resulting corridor or subarea study. The scope of that additional work will be determined during the NEPA process. If any additional reasonable alternatives are identified during project scoping they will also have to be studied during the NEPA process.

Of course, planning studies may lead to a class of action other than an EIS. As noted earlier, one of the advantages of corridor and subarea studies is that this pre-NEPA analysis can sometimes result in project decisions that may have cost and time savings, which may include avoidance of significant impacts. For an example of where a planning study led to a class of action other than an EIS, see Appendix B.1, the Libby North Corridor Study done by Montana Department of Transportation.

> **North-South Corridor Study**
> The Arizona Department of Transportation (ADOT) is leading an engineering and environmental study to identify and evaluate possible transportation routes for a proposed North-South Corridor between U.S. 60 and Interstate-10 near Florence, Arizona. Arizona DOT is relying on over 10 prior planning studies, and the extensive public/agency input obtained through those previous studies, for the North-South Corridor Study.
> http://www.azdot.gov/Highways/Projects/NorthSouthCorridorStudy/Index.asp

3.3.2 Using the Notice of Intent to Link Planning and NEPA

When an EIS is prepared, the connection between planning and NEPA can be made through the NOI. The NOI is published in the *Federal Register* by the lead Federal agency and announces an agency's decision to prepare an EIS for a particular action. The NOI describes the proposed action, possible alternatives, and the agency's proposed scoping process, as well as a point of contact. The NOI triggers the scoping process. To achieve linkage between planning and NEPA work, the NOI should refer to the relevant planning information that the lead agencies propose to use in NEPA, such as the preliminary purpose and need or the range of alternatives studied. These planning level studies and decisions by State or local entities have to be presented to the public and to the agencies involved for their input before the joint lead agencies make a decision on the purpose and need and the range of alternatives.[42]

---

[42] 23 U.S.C. § 139(f)(1) and (f)(4)(A).

The NOI language should clearly state whether the joint lead agencies propose to use analysis from any prior planning studies in the NEPA evaluation, and include the source of that analysis and where it is publicly available.

Examples of Notices of Intent that Link Planning and NEPA

Transportation Expansion (T-REX) Multi-Modal Transportation Project (Colorado)
"The FHWA and FTA are jointly issuing this notice to advise the public that an environmental impact statement will be prepared for the proposed transportation improvements in the Southeast Corridor of the Denver metropolitan area.... The proposed action is consistent with the recently completed Southeast Corridor Major Investment Study.... Transit and highway improvements are intended to alleviate traffic congestion in the Southeast Corridor, address safety problems and help achieve regional air quality goals by providing an alternative to the single occupant vehicle."
Reference: Federal Register / Vol. 63, No. 28 / Wednesday, February 11, 1998 (ROD signed March 2000)

I-95 Improvement Project (Connecticut)
"Improvements to the I-95 corridor are considered necessary to improve safety and to provide for increases in projected traffic volumes. Alternatives under consideration include, but are not limited to: (1) taking no action and (2) addition of a third travel lane in each direction. The EIS will use data and findings from two major deficiency and needs studies entitled "Southeastern Connecticut Corridor Study" dated January 1999 and "I-95 Corridor Feasibility Study, Branford to Rhode Island" dated December 2004. Copies of these studies are available from ConnDOT's Office of Environmental Planning."
Reference: Federal Register / Vol. 72, No. 162 / Wednesday, August 22, 2007

South Capitol Street Roadway Improvement and Bridge Replacement Project (Washington, DC)
"The project includes the proposed redevelopment of South Capitol Street per, the National Capital Planning Commission's 1997 plan, Extending the Legacy, Planning America's Capital for the 21st Century."
Reference: Federal Register / Vol. 70, No. 79 / Tuesday, April 26, 2005

Intercounty Connector (ICC) Project (Maryland)
"Project studies pursuant to the National Environmental Policy Act (NEPA) concerning the ICC project were most recently conducted in the early to late-1990s resulting in the completion of a Draft EIS/Draft Section 4(f) Evaluation in 1997. Study alternatives were presented at four Location/Design Public Hearings in May and June 1997. The State of Maryland put the ICC project on hold shortly after the hearings. The ICC project will involve the consideration of a reasonable range of alternatives that address the project goals. Consistent with NEPA, a full range of multi-modal highway alternatives will be considered, ranging from a No-Action Alternative to a limited access roadway on new location."
Reference: Federal Register / Vol. 68, No. 106 /Tuesday, June 3, 2003

4.0 Making a Planning Study Viable for NEPA

The previous section focused on the content of planning studies and considerations for transportation planners to conduct a study in accordance with regulations on the use of corridor and subarea studies in NEPA. This section provides guidance from the perspective of the NEPA practitioner on how transportation planners can best ensure that a corridor or subarea study can inform NEPA. If transportation planners prepare a study, keeping in mind the responsibilities of the NEPA practitioner, they can improve the viability of their planning documents during the NEPA process.

4.1 Weight Given to Planning Products Informing NEPA

While State and local entities are responsible for determining the long-term planning goals for their jurisdictions, including transportation goals[43], FHWA is ultimately responsible for ensuring NEPA compliance of transportation projects and therefore will make the final determination whether the planning products have a rational basis and accurate data and analyses to support decisions during NEPA.[44] As a threshold matter, the study must meet the regulatory criteria for use of a corridor or sub-area study in NEPA, as required by 23 CFR §§ 450.212(b) and 450.318(b). To determine the weight such a study can be given, these agencies will review the documentation to determine whether the data and analyses are reliable, current, and defensible. The data and analyses must meet requirements for professional integrity, including scientific integrity, of the decisions and analyses in environmental documents.[45] This review is part of the "hard look" requirement agencies must meet under NEPA.[46] If a study is not able to meet these integrity thresholds, it should be rejected for use in subsequent environmental documentation.[47]

4.2 Appropriate Environmental Analysis

Often, transportation planners look for information and guidance about the environmental context by engaging resource agencies. However, the environmental analysis found in corridor and subarea studies may not be as refined as what resource agencies are accustomed to in NEPA-quality, project-level environmental documentation. Transportation planners may be satisfied to get GIS files for specific populations, critical habitats, or other resources, and then 'layer' them together to get an overview of potential impacts or areas to avoid. Resource agencies that have not previously been engaged during the planning process may be hesitant to share data and actively participate in studies if rigorous scientific review or field inspections are not being conducted. When seeking resource and regulatory agency input to planning, it is generally more effective to engage directly with the agencies, such as by inviting agency staff to transportation staff meetings where decisions are being discussed, rather than providing partner resource agencies a planning document to review at the end of the process.

---

[43] *See, e.g., Citizens for Smart Growth v. Peters*, 716 F.Supp.2d 1215, 1223-1225 (S.D. Fla. 2010).
[44] 40 CFR 1506.5(a).
[45] 40 CFR 1502.24.
[46] *See Theodore Roosevelt Conservation Partnership v. Salazar*, 616 F.3d 497, 510-511 (D.C. Cir. 2010) (agency's substantive review of outdated methodology for estimating ozone impacts of oil and gas drilling met NEPA "hard look" requirement and Administrative Procedure Act "arbitrary and capricious" standard).
[47] *See, e.g., Utahns for Better Transp. v. U.S. Dept. of Transp.*, 305 F.2d 1152, *1181-1182 (10th Cir.2002)* (agencies' failure to verify applicant cost estimates failed to meet NEPA requirement that agencies ensure accuracy of information supplied by applicants); *see also, Citizens for Smart Growth v. Peters*, 716 F.Supp.2d 1215, 1225-1226 (S.D. Florida, 2010).

4.3 GOOD DOCUMENTATION

It is essential to document pre-NEPA analysis and decisions if an agency wants to use corridor and subarea planning studies to inform NEPA. Good documentation includes:

- Explaining the thought process underlying analytical conclusions and planning decisions, particularly when alternatives are analyzed and screened or eliminated;

- Describing the information used at the planning stage, including what that information is, how current or complete it is, and how reliable it is over time; and

- Documenting public and agency involvement.

The most robust documentation should be provided where the goal is to persuade joint lead agencies and others to adopt decisions made in the planning process, such as the identification of a range of alternatives for detailed analysis in NEPA. In this case, documentation must meet NEPA requirements if it is to be used in NEPA without additional analysis by the lead agencies. Transportation NEPA practitioners, resource agencies, and transportation planners should communicate to make sure there is mutual understanding of the documentation standards required. Transportation planners and NEPA practitioners should agree on the acceptable level of effort and documentation. The key to making the planning and environment linkage work is for transportation planners and NEPA practitioners to collaborate and develop agreed-upon documentation standards. NEPA practitioners have to be confident that the information they are receiving is valid and useful, if it is to become a basis for NEPA decision-making and part of the project's administrative record.

Examples of appropriate documentation tools in current use are provided in the subsections below.

4.3.1 Planning/ Environmental Linkages Questionnaire

FHWA recommends documenting planning-level analysis that can be used to inform NEPA. One tool to accomplish this is the Planning/Environmental Linkages Questionnaire. The questionnaire is intended to:

- Inform planners about the requirements and options to consider while developing a planning study with a goal to inform the NEPA process; and

- Document and share relevant planning information with NEPA practitioners to build understanding about a project – both the information studied and areas that require more analysis.

Once it is completed, the Questionnaire acts as a summary of the planning process and it eases the transition from planning to NEPA. The questionnaire is an adaptation of one developed by the Colorado DOT and FHWA Colorado Division Office. FHWA recommends that the following questions be used as a guide throughout the planning process, not answered at completion of the process. The questionnaire can be included in the planning document as an executive summary, chapter, or appendix.

Federal Highway Administration
Planning/Environmental Linkages Questionnaire

This questionnaire is intended to act as a summary of the Planning process and ease the transition from planning to a National Environmental Policy Act (NEPA) analysis. Often, there is no overlap in personnel between the planning and NEPA phases of a project, so consequently much (or all) of the history of decisions made in the planning phase is lost. Different planning processes take projects through analysis at different levels of detail. NEPA project teams may not be aware of relevant planning information and may re-do work that has already been done. This questionnaire is consistent with the 23 CFR 450 (Planning regulations) and other FHWA policy on Planning and Environmental Linkage (PEL) process.

The Planning and Environmental Linkages study (PEL Study) is used in this questionnaire as a generic term to mean any type of planning study conducted at the corridor or subarea level which is more focused than studies at the regional or system planning levels. Many States may use other terminology to define studies of this type and those are considered to have the same meaning as a PEL study.

At the inception of the PEL study, the study team should decide how the work may later be incorporated into subsequent NEPA efforts. A key consideration is whether the PEL study will meet standards established by NEPA regulations and guidance. One example is the use of terminology consistent with NEPA vocabulary (e.g. purpose and need, alternatives, affected environment, environmental consequences).

Instructions: These questions should be used as a guide throughout the planning process, not just answered near completion of the process. When a PEL study is started, this questionnaire will be given to the project team. Some of the basic questions to consider are: "What did you do?", "What didn't you do?" and "Why?". When the team submits a PEL study to FHWA for review, the completed questionnaire will be included with the submittal. FHWA will use this questionnaire to assist it in determining if the study meets the requirements of 23 CFR §§ 450.212 or 450.318. The questionnaire should be included in the planning document as an executive summary, chapter, or appendix.

1. Background:
    a. Who is the sponsor of the PEL study? (State DOT, Local Agency, Other)
    b. What is the name of the PEL study document and other identifying project information (e.g. sub-account or STIP numbers, long-range plan or transportation improvement program years)?
    c. Who was included on the study team (Name and title of agency representatives, consultants, etc.)?
    d. Provide a description of the existing transportation facility within the corridor, including project limits, modes, functional classification, number of lanes, shoulder width, access control and type of surrounding environment (urban vs. rural, residential vs. commercial, etc.)

      e. Provide a brief chronology of the planning activities (PEL study) including the year(s) the studies were completed.
      f. Are there recent, current or near future planning studies or projects in the vicinity? What is the relationship of this project to those studies/projects?

2. Methodology used:
    a. What was the scope of the PEL study and the reason for completing it?
    b. Did you use NEPA-like language? Why or why not?
    c. What were the actual terms used and how did you define them? (Provide examples or list)
    d. How do you see these terms being used in NEPA documents?
    e. What were the key steps and coordination points in the PEL decision-making process? Who were the decision-makers and who else participated in those key steps? For example, for the corridor vision, the decision was made by State DOT and the local agency, with buy-in from FHWA, the USACE, and USFWS and other resource/regulatory agencies.
    f. How should the PEL information be presented in NEPA?

3. Agency coordination:
    a. Provide a synopsis of coordination with Federal, tribal, State and local environmental, regulatory and resource agencies. Describe their level of participation and how you coordinated with them.
    b. What transportation agencies (e.g. for adjacent jurisdictions) did you coordinate with or were involved during the PEL study?
    c. What steps will need to be taken with each agency during NEPA scoping?

4. Public coordination:
    a. Provide a synopsis of your coordination efforts with the public and stakeholders.

5. Purpose and Need for the PEL study:
    a. What was the scope of the PEL study and the reason for completing it?
    b. Provide the purpose and need statement, or the corridor vision and transportation goals and objectives to realize that vision.
    c. What steps will need to be taken during the NEPA process to make this a project-level purpose and need statement?

6. Range of alternatives: Planning teams need to be cautious during the alternative screen process; alternative screening should focus on purpose and need/corridor vision, fatal flaw analysis and possibly mode selection. This may help minimize problems during discussions with resource agencies. Alternatives that have fatal flaws or do not meet the purpose and need/corridor vision will not be considered reasonable alternatives, even if they reduce impacts to a particular resource. Detail the range of alternatives considered, screening criteria and screening process, including:
    a. What types of alternatives were looked at? (Provide a one or two sentence summary and reference document.)

    b. How did you select the screening criteria and screening process?
    c. For alternative(s) that were screened out, briefly summarize the reasons for eliminating the alternative(s). (During the initial screenings, this generally will focus on fatal flaws)
    d. Which alternatives should be brought forward into NEPA and why?
    e. Did the public, stakeholders, and agencies have an opportunity to comment during this process?
    f. Were there unresolved issues with the public, stakeholders and/or agencies?

7. Planning assumptions and analytical methods:
    a. What is the forecast year used in the PEL study?
    b. What method was used for forecasting traffic volumes?
    c. Are the planning assumptions and the corridor vision/purpose and need statement consistent with each other and with the long-range transportation plan? Are the assumptions still valid?
    d. What were the future year policy and/or data assumptions used in the transportation planning process related to land use, economic development, transportation costs and network expansion?

8. Environmental resources (wetlands, cultural, etc.) reviewed. For each resource or group of resources reviewed, provide the following:
    a. In the PEL study, at what level of detail was the resource reviewed and what was the method of review?
    b. Is this resource present in the area and what is the existing environmental condition for this resource?
    c. What are the issues that need to be considered during NEPA, including potential resource impacts and potential mitigation requirements (if known)?
    d. How will the planning data provided need to be supplemented during NEPA?

9. List environmental resources you are aware of that were not reviewed in the PEL study and why? Indicate whether or not they will need to be reviewed in NEPA and explain why.

10. Were cumulative impacts considered in the PEL study? If yes, provide the information or reference where the analysis can be found.

11. Describe any mitigation strategies discussed at the planning level that should be analyzed during NEPA.

12. What needs to be done during NEPA to make information from the PEL study available to the agencies and the public? Are there PEL study products which can be used or provided to agencies or the public during the NEPA scoping process?

13. Are there any other issues a future project team should be aware of?
    a. Examples: Controversy, utility problems, access or ROW issues, encroachments into ROW, problematic land owners and/or groups, contact information for stakeholders, special or unique resources in the area, etc.

4.3.2 Corridor Planning Study Checklist

The State of Montana developed a checklist to allow planning studies to be moved into NEPA. The checklist is provided below and in the *Montana Business Process to Link Planning Studies and NEPA/MEPA Reviews.*[48] The checklist can be used as guidance at the beginning of and throughout the corridor planning process, and for confirmation at the end of the study. Montana intends to use it as an integral part of its overall business process in linking planning studies and environmental review.

---

[48] http://www.mdt.mt.gov/publications/docs/brochures/corridor_study_process.pdf.

Montana Department of Transportation
Corridor Planning Study Checklist

Introduction: Introductory information documenting:
- ☐ Identification of the Corridor Planning Study candidate;
- ☐ Reason(s) to conduct corridor planning;
- ☐ Study area definition (include map of the corridor boundaries and study area);
- ☐ General goals, objectives, and purpose of the study; and
- ☐ Members of the Corridor Planning Team.

Documentation and information from development of the work plan can be incorporated here.

Background: Background information on the corridor documenting:
- ☐ A summary of the review and documentation of previously developed information on conditions in the study corridor. Information gathered as part of the Corridor Setting Document may be used here.
- ☐ A summary of existing conditions in the study corridor. Detailed information, analysis, and results may be documented with Technical Reports and Data.

Identified Corridor Needs and Issues: Explain identified corridor needs and issues, documenting:
- ☐ Previously developed corridor needs, issues, and goals;
- ☐ Known corridor needs and issues; and
- ☐ Input from public involvement and resource and other agency consultation.

Information presented here can be used in developing the draft statement of purpose and need.

Public Involvement and Resource and Other Agency Consultation: Provide documentation of how and when the public involvement and resource and other agency consultation was conducted and completed. This can be documented as a summary of what occurred with detailed information included in an appendix or a technical report. Information from the Public Involvement Plan may be used here. Documentation should include the following:

Public Involvement
- ☐ How many and when public meeting were held;
- ☐ Newsletters, press releases, presentation materials, sign-in sheets, minutes, and summary of discussion and comments at public meetings; and
- ☐ Documentation of any decision, findings, or commitments at public meetings.

Resource and other Agency Consultation
- ☐ How and when resource and other agency consultation was conducted including coordination methods and contacts;
- ☐ The federal, tribal, state, and local agencies included; and
- ☐ Documentation of information gathered including attendance, issues, responses, decisions, resolutions, commitments, and concurrences.

Technical Reports and Data: Reports developed and used as part of the Corridor Planning Process should be summarized in the Corridor Study Report and included in the appendix. The types of reports should include: existing and projected conditions including social and economic, an environmental scan, design standards, corridor geometrics, traffic data, accident information, travel demand forecasting, and economic data. Other information may be included depending on the type of study. Information from the Existing and Projected Conditions Report may be used here. At a minimum, reports/data should include:
- ☐ Where information was derived, summary of analytical methods used, forecast information assumptions, projections, and data collection dates (maps, visual aids, and other graphics should be included for clarification);
- ☐ Description of findings, recommendations, and conclusions from previous studies and reports; and

- Sources for review and documentation include existing planning or engineering studies, land use plans, projects both initiated and complete, and other local planning documents appropriate for this study area. The report should reference sources of information.
- Information gathered may include transportation system conditions (roadway and multimodal operating conditions, safety, etc.), as well as land use, social, economic, and environmental conditions in the corridor.
- Any conclusions, recommendations, or action brought forward from previously developed documents or projects and considered for inclusion in the Corridor Planning Study.
    - Disclosure of missing or unavailable information.

Analysis Methods and Findings: Information from the technical repots/data and public/agency involvement to develop and eliminate alternatives. The section should include:
- Description of alternatives and/or options developed;
- Description of selection or screening criteria (this may include cost);
- Alternatives and/or options advanced and eliminated with a summary of the rationale; and
- Description of possible phasing of alternatives of interim solutions.

Funding: Description of funding scenarios. Include information documenting:
- Planning level cost estimates or projections for alternatives and/or options, both short and long term and phases;
- Concerns with funding of alternative(s) due to excessive cost;
- Sources and types of funding available including partnership opportunities with other agencies, private developers or other groups; and
- Funding challenges and possible solutions.

Summary/Recommendations: A summary of the Corridor Planning Process; the identified need, issues, and goals; the recommended alternatives and/or options to be carried forward; the draft statement of purpose and need; and an implementation strategy for moving to the project development stage should be documented.

Project Development: Documentation of the elements listed here should be developed and included in the Corridor Study Report or as a stand-alone report. These elements bring the Corridor Planning Study into project development. The following elements should be considered and documented:
- Describe which alternatives should be carried forward into a NEPA/MEPA study;
- Include any recommended coordination or steps to be taken with resource and other agencies during NEPA/MEPA process;
- Identify resource issues that need additional consideration and evaluation;
- Describe any additional data or gaps in data that must be supplemented during the NEPA/MEPA process;
- Describe any resources that were not reviewed and why;
- Forward any possible mitigation strategies (include avoidance); and
- Describe any other issues that should be brought to the attention of the future project team.

4.3.3 Issue/Concern Tracking and Response

The North Central Texas Council of Governments (NCTCOG) has been methodical in its documentation for the Regional Outer Loop Corridor Feasibility Study.[49] One tool it has used is a database that tracks public comments and responses. This helps preserve a project history in case there is staff turnover, while building the foundation for an administrative record for future environmental reviews. The database tracks the source and date of the comment, manner of communication, its particular focus, and NCTCOG response. An example of the database is presented below.

Commenter/ Agency	Commenter Name and Title	Comment	Date Received	Source	Category	Comment Response
Agency 1	Commenter 1	Why are the Outer Loop Study corridors so far from he center of the region?	6/18/2008	verbal	Alignments	Large lakes and existing developments are major obstacles that prevent the placement of corridors closer to the urban core.
Agency 2	Commenter 2	It will be important to iden ify the indirect impacts such as any economic development or sprawl associated with the project in the EIS documents.	6/18/2008	verbal	Development	Sprawl is a concern of NCTCOG. It should be possible to make design choices that reduce the amount of sprawl associated. Balancing the access and mobility requirements of the Outer Loop will require input from all stakeholders. Different portions of the Outer Loop may have different levels of access.
Agency 3	Commenter 3	What would happen in a situation where a city initially does not request access to the Outer Loop, but then changes heir decision?	6/18/2008	verbal	City Support	The Outer Loop Study process will work with all parties to determine the appropriate levels of access and that any changes made after the study is completed would need to be addressed in a new process.
Agency 4	Commenter 4	When will the transportation model be ready?	7/8/2008	verbal	Model	The expanded model will be available no later than 2010.

4.4 NEPA PRACTITIONER'S PERSPECTIVE

The statewide and metropolitan transportation planning process takes considerable time and involves numerous individuals, agencies, and stakeholder groups. From the initial steps of planning, through project development, the potential exists for staff turnover. Often, the individuals instrumental to corridor or subarea planning decisions are unavailable when project-level environmental review begins. Good documentation chronicling the State and local decisions made within planning can help avoid a scenario where analyses or decisions are unnecessarily revisited when the proposed project reaches the environmental review stage.

Even where there is no staff turnover, typically different staff are involved at different stages of the transportation planning and project development process. Transportation planners who worked on the long-range plan and subsequent studies may not be the same people managing a particular project through to development. Often, communication challenges across divisions and agencies exist. NEPA practitioners are often unaware of prior planning decisions and unable to draw from the analysis of previous corridor and subarea studies. To remedy this, it makes sense for NEPA practitioners to work with transportation planners to identify the planning documents relevant to a particular project and then review this prior planning documentation to see to what extent it is useful. Wording can be added to contractual agreements for NEPA analysis and project

---

[49] http://www.nctcog.dst.tx.us/index.asp.

development to require review and consideration of prior planning studies. Reviewing prior planning documentation has the potential to save costs, reduce litigation risk, and lead to improved collaborative relationships between transportation planners and NEPA practitioners.

5.0 Lessons Learned

States and metropolitan areas across the country have taken several different approaches to the use of corridor and subarea planning studies in informing NEPA. The traditional approach of making planning decisions under the formal NEPA umbrella may still hold for many regions. Partner agencies that may not have engaged in a corridor planning process may see the issuance of the NOI as indication they need to be involved. It may then be easier to bring stakeholders to the table.

Other regions are using pre-NEPA analysis to inform the environmental review process. A major motivator for these regions is they have seen how corridor studies can lower environmental review and project development costs. Costs may be lower because the studies support eliminating unreasonable alternatives from detailed NEPA analysis. With fewer alternatives to analyze in an EIS, an agency can save money. In addition, a planned project can be scaled to a level consistent with local needs. Most importantly, State and local agencies can identify fatal flaws in planning prior to entering NEPA. This can save considerable expense. If public opposition is found during planning, State and local agencies can identify strategies to address those concerns during planning, thereby possibly reducing environmental review costs and potential litigation further downstream. For an illustration of where a corridor study led to cost savings, see the case study on the Libby North Corridor Study (Appendix B.1).

Some regions have come to rely on planning studies because they have witnessed the time savings offered. Agencies that closely documented their planning process and resulting decisions have found that subsequent projects clear environmental review with minimal backtracking. Early involvement of resource agencies within transportation planning allows for a better appreciation of the transportation planning process and the ability of planning decisions to avoid potential environmental issues. This understanding can speed later environmental review of these decisions in NEPA. For an illustration of where planning-level analysis in a corridor study resulted in a starting point for staff entering future NEPA studies, see the case study on the Parker Road Corridor Study (Appendix B.2).

Agencies have learned that planning studies make sense for large projects or for projects with construction phasing. For large projects, planning studies can be particularly helpful because they look at the transportation need on a corridor or subarea level as opposed to analysis based on a specific alignment. This helps an agency clarify expectations and can lead to improved quality of decisionmaking when the need is viewed on a larger scale with the potential effects preliminarily identified. Large projects may also have phasing of individual improvements that have logical termini and independent utility, some 20 years or more into the future. For these projects, planning studies help preserve a project's history in case there is staff turnover, as well as build an administrative record for future environmental reviews. Agencies can use planning studies as a history of alternatives development, analysis, comments, and recommendations. By recording these decisions, their underlying reasons, and working with partner agencies and the public during the corridor planning study process, a State or local agency reduces the likelihood of revisiting planning decisions in future NEPA processes. For an illustration of where a planning study proved valuable to a large, multi-year project, see the case study on the Regional Outer Loop Feasibility Study (Appendix B.3).

6.0 USING TIERING TO CONNECT PLANNING WITH NEPA

In addition to using corridor and subarea planning to inform NEPA, agencies may choose to use a tiered EIS approach in their decisionmaking. Tiering refers to the process of addressing a broad, general program, policy or proposal in an initial EIS, and analyzing a narrower site-specific proposal, related to the initial program, policy, or proposal in a subsequent NEPA document. The CEQ regulations (40 CFR Parts 1500-1508) recognize the use of tiering as one option for complying with NEPA, as do FHWA regulations (23 CFR § 771.111(g)). The intent in tiering is to encourage agencies to eliminate repetitive discussions and also to focus on the actual issues which are ripe for decision at each level of environmental review. If tiering is utilized, the site-specific NEPA document (Tier 2) contains a summary of the issues discussed and any decisions made in the first document (Tier 1) and the agency will incorporate by reference discussions from the first document. Thus, the second or site-specific document would focus primarily on the issues relevant to the specific proposal, and would not duplicate material found in the first document. In these cases, there are some considerations to keep in mind.[50]

Tiering may be used to authorize corridor preservation when construction is several years away, The Tier 1 EIS has similarities to corridor and subarea planning in that it takes a higher level perspective and may be prepared well in advance of actual project construction.

Tiering requires extensive efforts to educate and explain the tiered process to agencies and the public. It may be difficult to reach agreement among agencies with jurisdiction on how to handle non-NEPA requirements e.g., Section 404 (discharge of dredge or fill material into water) permitting or Section 106 (historic properties) consultation in a tiered NEPA process.

Like corridor and subarea planning, tiering can be a useful method of reducing paperwork and duplication when used carefully for appropriate types of plans, programs, and policies, which will later be translated into site-specific projects.

---

[50] As taken from PB Americas, Inc. and Perkins Coie LLP, "Guidelines on the Use of Tiered Environmental Impact Statements for Transportation Projects." June 2009. Prepared as part of NCHRP Project 25-25, Task 38.

7.0 CONCLUSION

Using corridor and subarea planning to inform NEPA has a number of benefits. It can result in lower costs and time savings for transportation agencies, and lead to a more integrated planning process. Federal regulations permit the use of corridor and subarea planning to inform NEPA; yet, many regions around the country are reluctant to take advantage of the practice. This document responds to the need for additional guidance on how best to use corridor and subarea planning to link the transportation planning and NEPA processes.

There is good reason for both transportation planners and NEPA practitioners to use corridor and subarea studies to inform NEPA. For transportation planners at State DOTs and MPOs, it can make sense to conduct a study when there are limited resources (e.g., funding, staff resources) that make a full-scale NEPA evaluation problematic. For NEPA practitioners, corridor and subarea planning can lead to improved environmental documentation, maximize avoidance of impacts, and result in improved relationships with planning staffs.

When conducting corridor and subarea plans, particularly to inform NEPA, it is important to involve a broad range of partners, including resource and regulatory agencies, transportation NEPA practitioners, planning and development partners, legal counsel, and the public. It can be difficult to engage all and maintain involvement from these parties, but early and continuing engagement with stakeholders leads to better information and ultimately improved transportation decisionmaking.

The transportation planning regulations governing the use of transportation planning materials to inform project development identify products that corridor or subarea studies may produce for a proposed transportation project. In addition, the regulations lay out the conditions that must be met in order to use planning materials in a NEPA review.[51] Besides this regulatory information, those preparing a corridor or subarea study should keep in mind the NEPA practitioner's responsibilities, so as to improve the usefulness of their planning documents during NEPA. The most important of these responsibilities is good documentation that explains the thought process underlying planning decisions. Examples of good documentation, like the Planning/ Environmental Linkages Questionnaire and the Corridor Planning Study Checklist, are being used today by State DOTs and MPOs. By taking advantage of these resources, State and local agencies need not reinvent the wheel.

Corridor and subarea studies are not the only approach to link planning and NEPA, but in our experience they provide substantial benefits. Corridor and subarea studies can help agencies identify meaningful issues that should be carried over to NEPA; they enhance flexibility, build understanding between agencies, and respond to fiscal challenges. There is no guarantee that what is decided by State and local entities in corridor/subarea planning will be accepted by the lead Federal agency and stakeholders during NEPA, but the opportunity exists that identified problems can be addressed early and initial project expectations understood throughout the transportation planning and project development process.

---

[51] 23 CFR §§ 450.212(b) and 450.318(b).

Appendix A: Legal, Policy and Guidance Framework

A.1 NEPA and Transportation Decisionmaking

The U.S. Congress approved the National Environmental Policy Act of 1969, as amended (42 U.S.C. §§ 4321-4347) to establish a national policy to protect the environment. NEPA also authorized the Council on Environmental Quality (CEQ), which adopted regulations for implementing NEPA, including the classes of actions and the analyses required. NEPA provides a framework for transportation decisionmaking. The principles or essential elements of NEPA decisionmaking include:[52]

- Assessment of the social, economic, and environmental impacts of a proposed action or project
- Analysis of a range of reasonable alternatives to the proposed project, based on the applicant's defined purpose and need for the project
- Consideration of appropriate impact mitigation: avoidance, minimization, and compensation
- Interagency participation: coordination and consultation
- Public involvement including opportunities to participate and comment
- Documentation and disclosure.

The NEPA process begins with the establishment of the purpose and need for the project. Purpose and need is the foundation of the NEPA process. It establishes the reasons for proposing an action. The planning process can go a long way in establishing the purpose and need for the project with the input of the public and resource agencies.

Following the requirement to establish the purpose and need is the identification of logical termini for the project,[53] and development of a full range of alternatives to address the transportation and environmental performance measures and community goals. These alternatives are subsequently screened based on performance measures which can include: ability to meet project purpose and need, environmental impacts/mitigation, other project/community goals and objectives, as well as the ability to meet requirements of other statutes and regulations (e.g., Section 4(f), Section 106, Section 404, Section 7, Clean Air Act). Again, coordination with the public and resource agencies (project stakeholders) during the planning process can help to narrow down the alternatives evaluated in the NEPA process and to focus the NEPA document on the issues that are most important to the environment and community/stakeholders. Planning studies can help determine reasonable modal options and corridors, narrowing the focus for the NEPA studies.

Throughout the environmental analyses, there is a responsibility to coordinate with and engage the public and the agencies (Cooperating Agencies, Participating Agencies). Where the public and agencies have been involved in the planning process to establish project needs and help determine the range of alternatives to be studied, these involvement activities, where adequately documented, can be referenced and summarized to carry forward into the NEPA process in accordance with applicable regulations. .

---

[52] http://www.environment.fhwa.dot.gov/projdev/pd3tdm.asp
[53] 23 CFR § 771.111(f).

NEPA is highly "procedural"; therefore documentation of the process that was followed is a necessity. The NEPA decision is based on the studies conducted and information collected.

A.2 FHWA/FTA REGULATORY LANGUAGE (AS OF THE DATE OF THIS GUIDANCE)

A.2.1 40 CFR Part 1500, CEQ Regulations for Implementing NEPA

To assist Federal agencies in effectively implementing the environmental policy and "action forcing" provisions of NEPA, the CEQ issued Regulations for Implementing the Procedural Provisions of the National Environmental Policy Act (40 CFR Parts 1500-1508). The regulations are applicable to and binding on all Federal agencies for implementing the provisions of NEPA except where compliance would be inconsistent with other statutory requirements. The CEQ regulations for implementing NEPA are available at: http://ceq.hss.doe.gov/nepa/regs/ceq/toc_ceq.htm

A.2.2 23 CFR Part 450 Statewide Transportation Planning; Metropolitan Transportation Planning

On February 14, 2007, FHWA and FTA issued new planning regulations that eliminated the requirement for a major investment study and implemented provisions enacted by SAFETEA-LU. In its place, the regulations created a new optional procedure for linking transportation planning and NEPA studies. The procedures are contained in 23 CFR § 450.212 (statewide planning) and § 450.318 (metropolitan planning). The full text is provided below.

§ 450.212 Transportation planning studies and project development.

(a) Pursuant to section 1308 of the Transportation Equity Act for the 21st Century, TEA–21 (Pub. L. 105–178), a State(s), MPO(s), or public transportation operator(s) may undertake a multimodal, systems-level corridor or subarea planning study as part of the statewide transportation planning process. To the extent practicable, development of these transportation planning studies shall involve consultation with, or joint efforts among, the State(s), MPO(s), and/or public transportation operator(s). The results or decisions of these transportation planning studies may be used as part of the overall project development process consistent with the National Environmental Policy Act (NEPA) of 1969 (42 U.S.C. 4321 *et seq.*) and associated implementing regulations (23 CFR part 771 and 40 CFR parts 1500–1508). Specifically, these corridor or subarea studies may result in producing any of the following for a proposed transportation project:

(1) Purpose and need or goals and objective statement(s);

(2) General travel corridor and/or general mode(s) definition (e.g., highway, transit, or a highway/transit combination);

(3) Preliminary screening of alternatives and elimination of unreasonable alternatives;

(4) Basic description of the environmental setting; and/or

(5) Preliminary identification of environmental impacts and environmental mitigation.

(b) Publicly available documents or other source material produced by, or in support of, the transportation planning process described in this subpart may be incorporated directly or by reference into subsequent NEPA documents, in accordance with 40 CFR 1502.21, if:

(1) The NEPA lead agencies agree that such incorporation will aid in establishing or evaluating the purpose and need for the Federal action, reasonable alternatives, cumulative or other impacts on the human and natural environment, or mitigation of these impacts; and

(2) The systems-level, corridor, or subarea planning study is conducted with:

(i) Involvement of interested State, local, Tribal, and Federal agencies;

(ii) Public review;

(iii) Reasonable opportunity to comment during the statewide transportation planning process and development of the corridor or subarea planning study;

(iv) Documentation of relevant decisions in a form that is identifiable and available for review during the NEPA scoping process and can be appended to or referenced in the NEPA document; and

(v) The review of the FHWA and the FTA, as appropriate.

(c) By agreement of the NEPA lead agencies, the above integration may be accomplished through tiering (as described in 40 CFR 1502.20), incorporating the subarea or corridor planning study into the draft Environmental Impact Statement or Environmental Assessment, or other means that the NEPA lead agencies deem appropriate. Additional information to further explain the linkages between the transportation planning and project development/NEPA processes is contained in Appendix A to this part, including an explanation that is non-binding guidance material.

§ 450.318 Transportation planning studies and project development.

(a) Pursuant to section 1308 of the Transportation Equity Act for the 21st Century, TEA–21 (Pub. L. 105–178), an MPO(s), State(s), or public transportation operator(s) may undertake a multimodal, systems-level corridor or subarea planning study as part of the metropolitan transportation planning process. To the extent practicable, development of these transportation planning studies shall involve consultation with, or joint efforts among, the MPO(s), State(s), and/or public transportation operator(s). The results or decisions of these transportation planning studies may be used as part of the overall project development process consistent with the National Environmental Policy Act (NEPA) of 1969 (42 U.S.C. 4321 *et seq.*) and associated implementing regulations (23 CFR part 771 and 40 CFR parts 1500–1508). Specifically, these corridor or subarea studies may result in producing any of the following for a proposed transportation project:

(1) Purpose and need or goals and objective statement(s);

(2) General travel corridor and/or general mode(s) definition (e.g., highway, transit, or a highway/transit combination);

(3) Preliminary screening of alternatives and elimination of unreasonable alternatives;

(4) Basic description of the environmental setting; and/or

(5) Preliminary identification of environmental impacts and environmental mitigation.

(b) Publicly available documents or other source material produced by, or in support of, the transportation planning process described in this subpart may be incorporated directly or by reference into subsequent NEPA documents, in accordance with 40 CFR 1502.21, if:

(1) The NEPA lead agencies agree that such incorporation will aid in establishing or evaluating the purpose and need for the Federal action, reasonable alternatives, cumulative or other impacts on the human and natural environment, or mitigation of these impacts; and

(2) The systems-level, corridor, or subarea planning study is conducted with:

(i) Involvement of interested State, local, Tribal, and Federal agencies;

(ii) Public review;

(iii) Reasonable opportunity to comment during the metropolitan transportation planning process and development of the corridor or subarea planning study;

(iv) Documentation of relevant decisions in a form that is identifiable and available for review during the NEPA scoping process and can be appended to or referenced in the NEPA document; and

(v) The review of the FHWA and the FTA, as appropriate.

(c) By agreement of the NEPA lead agencies, the above integration may be accomplished through tiering (as described in 40 CFR 1502.20), incorporating the subarea or corridor planning study into the draft Environmental Impact Statement (EIS) or Environmental Assessment, or other means that the NEPA lead agencies deem appropriate.

(d) For transit fixed guideway projects requiring an Alternatives Analysis (49 U.S.C. 5309(d) and (e)), the Alternatives Analysis described in 49 CFR part 611 constitutes the planning required by section 1308 of the TEA-21. The Alternatives Analysis may or may not be combined with the preparation of a NEPA document (e.g., a draft EIS). When an Alternatives Analysis is separate from the preparation of a NEPA document, the results of the Alternatives Analysis may be used during a subsequent environmental review process as described in paragraph (a).

(e) Additional information to further explain the linkages between the transportation planning and project development/NEPA processes is contained in Appendix A to this part, including an explanation that it is nonbinding guidance material.

A.2.3 23 CFR Part 771, Environmental Impact and Related Procedures

23 CFR Part 771 comprises the environmental regulations of the FHWA and FTA. In accordance with CEQ requirements, these regulations were adopted to implement NEPA requirements for highway and public transportation projects. Relevant portions concerning the use of corridor and subarea planning include:

§ 771.111 Early coordination, public involvement, and project development.

(a)(1) Early coordination with appropriate agencies and the public aids in determining the type of environmental review documents an action requires, the scope of the document, the level of analysis, and related environmental requirements. This involves the exchange of information from the inception of a proposal for action to preparation of the environmental review documents. Applicants intending to apply for funds should notify the Administration at the time that a project concept is identified. When requested, the Administration will advise the applicant, insofar as possible, of the probable class of action and related environmental laws and requirements and of the need for specific studies and findings which would normally be developed concurrently with the environmental review documents.

(2) The information and results produced by, or in support of, the transportation planning process may be incorporated into environmental review documents in accordance with 40 CFR 1502.21 and 23 CFR 450.212 or 450.318.

§ 771.123(b):

After publication of the Notice of Intent, the lead agencies, in cooperation with the applicant (if not a lead agency), will begin a scoping process which may take into account any planning work already accomplished, in accordance with 23 CFR 450.212 or 450.318. The scoping process will be used to identify the purpose and need, the range of alternatives and impacts, and the significant issues to be addressed in the EIS and to achieve the other objectives of 40 CFR 1501.7.

A.2.4 Appendix A to 23 CFR Part 450- Linking the Transportation Planning and NEPA Processes

On February 14, 2007, FHWA and FTA issued guidance on incorporating products of the planning process into NEPA documents as Appendix A of 23 CFR part 450. Appendix A uses a question and answer approach to discuss documentation needs and the level of detail of a planning product compared to a full NEPA analysis. The appendix outlines the type and extent of agency involvement in order for a planning process to be more readily accepted in NEPA review. The procedures for using decisions or analysis from the transportation planning process and methods for maximizing the likelihood that the results of planning activities will be adopted in the NEPA review are also provided. Appendix A is available on the FHWA Web site at
http://www.environment.fhwa.dot.gov/integ/publications.asp#exec

A.2.5 SAFETEA-LU (Public Law 109-59) Environmental Review Process FHWA/FTA Final Guidance, November 15, 2006

On November 15, 2006, FHWA and FTA issued joint guidance on the environmental review process required by Section 6002 of SAFETEA-LU. This section of SAFETEA-LU prescribes changes to existing FHWA and FTA procedures for implementing NEPA. The guidance provides explanations of new and changed aspects of the environmental review process for FHWA and FTA NEPA practitioners. It informs the reader about what, and how, things need to be done differently as a result of SAFETEA-LU. The guidance is divided into three sections: (1) environmental review process; (2) process management; and (3) statute of limitations. A question and answer format is used throughout the guidance. The final guidance is available on the FHWA Web site at
http://www.fhwa.dot.gov/hep/section6002/

A.2.6 FHWA Technical Advisory T 6640.8A, October 30, 1987 Guidance for Preparing and Processing Environmental and Section 4(f) Documents

In this technical advisory, FHWA provides guidance to FHWA field offices and to project applicants on the preparation and processing of environmental and Section 4(f) documents. Guidance is not regulatory but was developed to provide information to promote uniformity and consistency in the format, content, and processing of environmental documents. It should be used in combination with a knowledge and understanding of more recent Title 23 statutes, regulations and implementing guidance, the CEQ Regulations for Implementing NEPA (40 CFR Parts 1500-1508), FHWA's Environmental Impact and Related Procedures (23 CFR Part 771) and other environmental statutes, regulations, and orders. The technical advisory is available on the FHWA Web site at
http://www.environment.fhwa.dot.gov/projdev/impta6640.asp#aa

A.3 OTHER LAWS, REGULATIONS, AND ORDERS GOVERNING ENVIRONMENTAL DECISIONMAKING

The NEPA review process acts as an "umbrella" for compliance with a host of other Federal, State, and local requirements. Many of the specific resources that must be evaluated as part of the NEPA process also fall within the jurisdiction of other laws. It is important to ensure that these other requirements are coordinated and addressed. These other laws may require a specific standard be met (substantive requirements) and/or may require a permit or other approval to be issued by the agency/entity responsible for administering the law. A few of these other Federal laws, regulations, and orders governing environmental decisionmaking are:

Clean Air Act (42 U.S.C. §7401 et seq. (1970), as amended; implementing regulations are in 40 CFR Parts 51-99)

The Clean Air Act (CAA) is the comprehensive Federal law that regulates air emissions from stationary and mobile sources. Among other things, this law authorizes EPA to establish National Ambient Air Quality Standards (NAAQS) to protect public health and public welfare and to regulate emissions of hazardous air pollutants. Regulations on implementation plans (40 CFR Part 51) and transportation conformity (40 CFR Part 93) are particularly relevant.

Clean Water Act - Section 404 (33 U.S.C. §1251 et seq. (1972), as amended; implementing regulations are found in 33 CFR Part 325 and 40 CFR Part 230)

The Clean Water Act (CWA) establishes the basic structure for regulating discharges of pollutants into the waters of the United States and regulating quality standards for surface waters. Section 404 prohibits any filling of wetlands or other discharges into the waters of the US without a permit issued by the US Army Corps of Engineers, subject to the possible veto by the US EPA.

Endangered Species Act - (16 U.S.C. §1531 et seq. (1973); joint implementing regulations of the lead Federal agencies (US Fish & Wildlife Service and NOAA Fisheries) appear in 50 CFR Parts 401-453)

The Endangered Species Act (ESA) provides a program for the conservation of threatened and endangered plants and animals and the habitats in which they are found. The lead Federal agencies for implementing ESA are the U.S. Fish and Wildlife Service (FWS) and the U.S. National Oceanic and Atmospheric Administration (NOAA) Fisheries Service. The FWS maintains a worldwide list of endangered species. Species include birds, insects, fish, reptiles, mammals, crustaceans, flowers, grasses, and trees.

Section 7 of the Act requires Federal agencies, in consultation with the U.S. Fish and Wildlife Service and/or the NOAA Fisheries Service, to ensure that actions they authorize, fund, or carry out are not likely to jeopardize the continued existence of any listed species or result in the destruction or adverse modification of designated critical habitat of such species unless an exemption has been granted. The law also prohibits any action that causes a "taking" of any listed species of endangered fish or wildlife. Likewise, import, export, interstate, and foreign commerce of listed species are all generally prohibited.

National Historic Preservation Act - Section 106 (16 U.S.C. § 470f; implementing regulations are found in 36 CFR Part 800)

The National Historic Preservation Act (NHPA) (16 U.S.C. § 470 et seq.) is legislation intended to preserve historical and archaeological sites in the United States of America. The Act created the National Register of Historic Places, the list of National Historic Landmarks, and authorized criteria for allowing administration of the historic preservation program by State Historic Preservation Officers. Section 106 of the NHPA mandates Federal agencies undergo a review process for all federally-funded and federally-approved projects that will impact sites listed on, or eligible for listing on, the National Register of Historic Places. Specifically it requires the Federal agency to "take into account" the effect a project may have on historic properties. It allows interested parties an opportunity to comment on the potential impact projects may have on significant archaeological or historic sites. The main purpose for the establishment of the Section 106 review process is to minimize potential harm and damage to historic properties.

Department of Transportation Act - Section 4(f) (49 U.S.C. § 303, 23 U.S.C. § 138; FHWA/FTA implementing regulations are in 23 CFR Part 774)

The Department of Transportation Act (DOT Act) of 1966 included a special provision - Section 4(f) - which stipulated that the FHWA and other DOT agencies cannot approve the use of land from publicly owned parks, recreational areas, wildlife and waterfowl refuges, or public and private historical sites unless the following conditions apply: (1)There is no feasible and prudent alternative to the use of land and (2) The action includes all possible planning to minimize harm to the property resulting from use.

Executive Order 12898: Environmental Justice (59 FR 7629 (February 16, 1994); U.S. DOT's implementing Strategy is found at 60 FR 33896 (June 21, 1995) and implementing Order is found at 62 FR 18377 (April 15, 1997))

Executive Order (EO) 12898 orders executive agencies, with respect to their programs, policies, and activities, to identify and address disproportionately high and adverse impacts on minority and low income populations with respect to human health and the environment.

Executive Order 11990: Protection of Wetlands (42 FR 26961 (May 24, 1977); implementing DOT Order 5660.1A (August 24, 1978) and implementing FHWA regulations are found in 23 CFR 777)

Executive Order (EO) 11990 orders executive agencies to avoid direct or indirect support of new construction in wetlands wherever there is a practicable alternative.

State Laws -

Many States have their own State "mini-NEPA" laws and State transportation acts that require environmental reviews of State actions. Agencies should consider this requirement when preparing their corridor and subarea studies.

A.4 CURRENT POLICY FRAMEWORK

The FHWA encourage the use of corridor and subarea planning to inform NEPA.[54] Incorporating planning analysis into NEPA has several benefits. Corridor and subarea studies allow for environmental input and 'shaping' of early transportation planning decisions by all involved agencies and the public. This should result in more informed planning decisions and could provide a level of predictability that proves advantageous, as subsequent environmental reviewers would be familiar with this earlier work. Making planning decisions, such as the nature of the transportation need, during planning, rather than during NEPA, can shorten project delivery. Corridor and subarea study procedures encourage interagency cooperation. Pre-NEPA analysis can enhance the quality of a project and reduce its impacts on the surrounding environment. The 2007 planning regulations (23 CFR Part 450) give authority to State DOTs and MPOs to link planning and NEPA, as do relevant NEPA implementing procedures (40 CFR Parts 1500-1508, 23 CFR Part 771). U.S. DOT encourages the full utilization of these authorities.

FHWA views corridor and subarea studies as one technique in helping prepare the highway community to meet the needs of the 21st century transportation system and economy. The use of planning studies to inform NEPA falls within the administration's Planning and Environment Linkages (PEL) initiative. PEL represents an approach to transportation decisionmaking that considers environmental, community, and economic goals early in the planning stage and carries them through project development, design, and construction. The goal of PEL is to create a

---

[54] 23 CFR 450 Appendix A.

seamless decisionmaking process that minimizes duplication of effort, promotes environmental stewardship, and reduces delays from planning to project implementation.

A.5 NCHRP REPORT 435: GUIDEBOOK FOR TRANSPORTATION CORRIDOR STUDIES

Relying on pre-NEPA analyses for environmental documentation is not a new concept. In 1999, the National Cooperative Highway Research Program (NCHRP) developed Report 435: "Guidebook for Transportation Corridor Studies: A Process for Effective Decisionmaking." The Guidebook documents research in the design and management of corridor and subarea transportation planning studies. It provides practical tools and guidance for designing, organizing, and managing these studies. The Guidebook brings together lessons learned from actual experiences in different regions of the country and on corridor/subarea studies with different scopes and levels of complexity. It provides a structured approach that State DOTs, MPOs, and local transportation planners can take to support transportation investments tailored to specific conditions and performance needs for major transportation improvements.

The Guidebook examines a corridor/subarea study as being primarily a decisionmaking process, as part of the stream of decisions that ultimately results in the implementation of transportation strategies that address an identified problem. The Guidebook does not cover all possible types of studies but focuses mainly on those involving potentially major infrastructure decisions.

There is no one size that fits all in the design and conduct of a corridor study, but the Guidebook follows the typical flow of a corridor study and is organized along the following chapters:

(1) Orientation to the Guidebook and Key Issues;

(2) The Transportation Planning Process and Corridor Decisionmaking;

(3) Identifying the Problem and the Corridor Study Strategy;

(4) Study Organization and Initiation;

(5) Community Involvement and Outreach;

(6) Confirming the Problem and Developing Evaluation Criteria;

(7) Developing and Evaluating Alternatives;

(8) Financial Analysis and Selection of the Preferred Investment Strategy;

(9) Corridor Study Documentation;

(10) Dealing with Technical and Institutional Issues That Arise During a Corridor Study; and

(11) Actions Agencies Can Take to Facilitate the Conduct of Corridor Planning Studies.

In addition to discussing the steps of the planning process for corridor studies, the Guidebook addresses the decisionmaking process and its relationship to NEPA. The Guidebook recommends development of core competencies to enable corridor studies. It recommends training for the following: modeling, public involvement and consensus building, economic analysis, financial analysis and funding. In sum, the Guidebook recommends that corridor studies can be made easier if agencies set the stage through their regular, ongoing planning activities. The development of some of these capabilities takes time, and an agency cannot expect these capabilities to be available unless it plans ahead and invests over the long term.

NCHRP Report 435 is available through the Transportation Research Board at
http://books.trbbookstore.org.

APPENDIX B: CASE STUDIES

B.1　LIBBY NORTH CORRIDOR STUDY

The Libby North Corridor Study is an evaluation by the Montana Department of Transportation (MDT) of an environmentally complex 14-mile section of Highway 567 abutting Pipe Creek in Kootenai National Forest in the Cabinet-Yaak Mountains of northwest Montana. Highway 567 runs between the City of Libby and the community of Yaak. The purpose of the study is to develop a comprehensive, long-range plan for managing and improving the corridor (known locally as Pipe Creek Road). Highway 567 is a two-lane roadway functionally classified as a rural major collector and is part of the Montana Secondary Highway System. The road provides access to U.S. Forest Service (USFS) lands for skiing, hunting, camping, and hiking activities and has historically been used for logging; that use continues today. The study evaluates existing conditions and determines what, if any, improvements should be made.

MDT options included a range of low-level safety type improvements through major reconstruction. Activities MDT undertook included researching existing conditions; documenting existing and projected environmental, geotechnical and land use conditions; forecasting future growth; identifying goals and analyzing alternatives for the corridor from several perspectives, constructability, financial feasibility, and public appearance; and recommending improvements and management strategies for the existing and long-term safety and operation of the corridor.

From the outset, MDT followed the transportation planning regulations as they pertained to the use of corridor studies. MDT held monthly meetings with USFS, FHWA, county staff, and elected officials. Recommendations from the meetings were reviewed by the agency and MDT staff for input and further collaboration. Once the range of recommendations was developed, continued meetings with the impacted resource agency staff allowed for discussion of concerns with the proposed build alternatives and improvement options. MDT's relationship with various resource agency staff improved through its outreach and collaboration, allowing for an inclusive determination of a solution.

Additional MDT outreach with the public brought all voices together to develop recommendations that could be forwarded within the environmental and funding constraints. Having public meetings to determine roadway concerns brought the public into the decisionmaking process. Early engagement of the public assisted in determining the most important concerns and needs for the roadway. Meeting these requirements assisted in the scoping of the project. Building better relationships with stakeholders also likely lessened conflicts in the future.

The Libby North Corridor Study saw a significant change in scope due to stakeholder discussions. The proposed project changed from full reconstruction to minor widening and alignment changes. Assessing the corridor at the planning level allowed for better understanding of corridor limitations and needs, and the improvements that could be reasonably pursued for this environmentally-sensitive corridor. With the substantially reduced impact to environmental, cultural, and social resources, the project shifted away from a costly EIS process to an anticipated Categorical Exclusion.

Lead Agency: Montana Department of Transportation
Partners: Lincoln County Commissioners, U.S. Forest Service, Federal Highway Administration
Internet Link: http://www.mdt.mt.gov/pubinvolve/libby

B.2 PARKER ROAD CORRIDOR STUDY

The Parker Road Corridor Study examined potential transportation solutions for the increasingly congested and rapidly growing Parker Road corridor between Hampden Avenue and E-470 in Denver, Colorado. Parker Road (SH 83) is a major north-south regional arterial with four to six through-lanes extending north into Denver and south into Colorado Springs. The eight-mile study corridor was a diagonal route from rapidly growing Douglas County northwest along Cherry Creek State Park into Denver. Current and projected traffic volumes and increasing traffic congestion along Parker Road prompted Arapahoe County to initiate a corridor transportation study to address regional mobility and local accessibility needs now and well into the future.

The study included several key elements in order to explore and define the most appropriate set of improvements to best move traffic through the corridor, while considering the unique context of this rapidly developing area. Elements included defining the immediate and long-term transportation needs of the corridor, as well as using Context Sensitive Design to look at roadway types and travel mode options. Recommendations included improvements to the roadway corridor, major intersections, bicycle/pedestrian facilities, transit services, and system management options.

Steps were taken to streamline the NEPA process for future improvements along the corridor. For instance, a vision statement and list of project objectives were developed in a format which could smoothly transition into a Purpose and Need statement; alternatives were evaluated thoroughly and screened through a specific process; public support and resource agency input was valued and requested throughout; and several aspects of the NEPA process received approval from the Executive Committee in charge at key decision points, including the range of alternatives, vision statement and objectives, major screening criteria, and recommended option.

Throughout the study, community and resource agency engagement were emphasized in order to gain as much input as possible. Meetings were held with State, Federal and regional resource and coordinating agencies early in the study to discuss corridor planning processes and potential resource impacts. Resource agency input was also requested regarding suggested "check-in" points, methods of documenting agency "buy-in," and the manner in which to pave the way for integrating planning study recommendations into NEPA without backtracking.

In addition to resource agency input, public opinion was also solicited. A Community Resource Panel (CRP) was formed to advise the project team of stakeholder concerns. The CRP was divided into four separate focus groups which included representatives from:
- Bicycle/Pedestrian advocates and groups
- Businesses, Metro Districts and Chambers of Commerce
- Emergency Services Providers
- Homeowners' Associations and Neighborhood Associations

The project team worked with each focus group throughout the project to identify project needs, review proposed alternatives, discuss potential impacts of improvements and possible mitigation techniques, and provide input on project implementation and phasing. The CRP also provided feedback on effective project communication tactics. Examples of outreach included news releases, local newspaper and television station advertisements, community bulletins, project update postcards via hard mail or e-mail, electronic mailing lists, and project newsletters.

Lead Agency: Arapahoe County (Colorado)
Partners: Colorado Department of Transportation, Federal Highway Administration, Denver Regional Council of Governments, Regional Transportation District, bordering cities, and towns
Internet Link: http://www.parkeronline.org/index.aspx?NID=571

B.3 REGIONAL OUTER LOOP CORRIDOR FEASIBILITY STUDY

The Regional Outer Loop Corridor Feasibility Study is an evaluation of the need and feasibility for an outer loop around the Dallas-Fort Worth region. This 240-mile regional Outer Loop would be a network of transportation routes that could incorporate existing and new highways, railways, and utility right-of-ways. The purpose of the study is to identify a single ½-to 1-mile wide Locally Preferred Corridor for the Regional Outer Loop and define Sections of Independent Utility.

From the outset, NCTCOG planned to integrate its studies into subsequent environmental documentation. The structure of the planning study follows a traditional NEPA document and includes a detailed history of the project and previous studies, need and purpose, existing and proposed conditions, development of alternatives, public and agency involvement and comments, and recommendations. The Appendix provided in the study helps to address the possibility of information becoming outdated over time. To aid subsequent studies in determining whether the study's resource data is useable, the Appendix creates a profile for each resource, detailing the source for the data and thoroughly documenting all of the information used as the basis for the recommendations. With this information, developers of future studies can understand why decisions were made, and outdated data sets can be identified, yielding potential time savings.

The feasibility study involves public and agency input throughout and based need for new transportation facilities on current population and projected growth within the region. NCTCOG has collected data for the existing and future conditions in the corridor and has had early and continuous information exchange with its partners to integrate environmental planning factors into all study phases. The "Bottom Up Approach" integrates stakeholder coordination effort through roundtables, public meetings, and briefings, thereby promoting buy-in early in the study phase and maintaining transparency.

One tool NCTCOG has found valuable in its corridor study has been GISST, a GIS-based tool that uses criteria that are scored to assist in initial assessment of potential environmental impacts. It helps decision makers better understand the potential significance of direct, indirect, and cumulative effects, and communicate with stakeholders about technical data. Another tool, NEPAssist, is a Web-based application which helps with the environmental review process and project planning. Drawing data dynamically from EPA GIS and other sources, it provides an initial environmental assessment of a project's potential environmental impacts, thereby raising issues early at the planning stage.

NCTCOG is still in the corridor evaluation phase and has not made any final recommendations, with the exception of a small length of road south east of Dallas.

Lead Agency: North Central Texas Council of Governments
Partners: U.S. Environmental Protection Agency, Federal Highway Administration, Texas Department of Transportation
Internet Link: http://www.nctcog.org/trans/spd/roadway/oloop/index.asp

APPENDIX C: RESOURCES

C.1 CORRIDOR AND SUBAREA PLANNING

- Transportation Research Board, "Guidebook for Transportation Corridor Studies: A Process for Effective Decision-Making," NCHRP Report 435, 1999.
    - The *Guidebook* documents research into the design and management of corridor and subarea transportation planning studies. It provides practical tools and guidance for designing, organizing, and managing these studies. It brings together lessons learned from actual experiences in different regions of the country and on corridor/subarea studies with different scopes and levels of complexity. It provides a structured approach that State DOTs, MPOs, and local transportation planners can take to support transportation investments tailored to specific conditions and performance needs for major transportation improvements.
    - http://books.trbbookstore.org

- Center for Environmental Excellence, "AASHTO Practitioner's Handbook # 10: Using the Transportation Planning Process to Support the NEPA Process," February 2008.
    - This handbook is intended to help transportation planners and National Environmental Policy Act practitioners improve linkages between the planning and NEPA processes, while also complying with recent legislative changes that require increased consideration of environmental issues in the planning process.
    - http://environment.transportation.org/pdf/programs/practitioners_handbook10.pdf

- Federal Highway Administration, "FHWA Peer Exchange on Using Corridor Planning to Inform NEPA," December 31, 2009.
    - This report summarizes an FHWA peer exchange, which examined the use of corridor planning studies as a foundation for NEPA decisionmaking. The report summarizes presentations of general trends as well as particular example projects.
    - http://www.environment.fhwa.dot.gov/integ/peer_exch_corridors.asp

C.2 FEDERAL REGULATIONS AND GUIDANCE

- U.S. Code of Federal Regulations "Title 23, Part 450 – Planning Assistance and Standards," 2007.
    - This rule governs highway planning, project development.
    - http://ecfr.gpoaccess.gov/cgi/t/text/text-idx?c=ecfr;sid=fa35bcb82ef73791b69348dc7d972318;rgn=div5;view=text;node=23%3A1.0.1.5.11;idno=23;cc=ecfr

- FHWA and FTA, "Guidance on Linking Transportation Planning and NEPA Documents," February 14, 2007 (Appendix A to 23 CFR 450).
    - This Appendix provides additional information to explain the linkage between the transportation planning and project development/NEPA processes. It is non-binding and should not be construed as a rule of general applicability.
    - http://edocket.access.gpo.gov/2007/pdf/07-493.pdf

- Schaftlein, Shari "Overview of SAFETEA-LU Sections 6001, 6002, 3005, and 3006," January 13, 2008.
    > This presentation explains several key provisions of SAFETEA-LU related to NEPA, planning, and decisionmaking.
    > http://www.environment.fhwa.dot.gov/integ/resources_publications_schaftlein.asp
- Federal Highway Administration, "Interagency Guidance: Transportation Funding for Federal Agency Coordination Associated with Environmental Streamlining Activities," 2007.
    > This report gives guidance to transportation and resource agencies looking for Federal funding to support resource agency collaboration and staff time in Planning and Environment Linkages.
    > http://www.environment.fhwa.dot.gov/integ/publications.asp
- U. S. Department of Transportation, "Environment and Planning Linkages Legal Guidance," 2005.
    > This memo pertains to the use of transportation planning documents and resources within the NEPA process, and the extent to which the two processes may rely on one another. The 2007 planning regulation amendments supersede the 2005 legal guidance if there is disagreement in content.
    > http://www.fhwa.dot.gov/hep/plannepalegal050222.htm
- Federal Highway Administration, "Every Day Counts (EDC) Initiative,"
    > This FHWA initiative promotes the rapid deployment of innovative technology to speed up project delivery while also enhancing safety and sustainability.
    > http://www.fhwa.dot.gov/publications/focus/10jun/02.cfm

C.3 PLANNING AND ENVIRONMENT LINKAGES

- National Cooperative Highway Research Project, "Guidance for Better Linking Systems Planning and the NEPA Process," January, 2006.
    > The publication profiles several examples of the incorporation of environmental factors into systems planning. Localities include Riverside County, CA and Tucson, AZ. The report identifies information and data needed and identifies barriers and recommendations for further integration of planning and NEPA processes.
    > http://144.171.11.40/cmsfeed/TRBNetProjectDisplay.asp?ProjectID=1262
- National Cooperative Highway Research Program "Consideration of Environmental Factors in Transportation," 2005.
    > This report describes the status of integrating environmental planning into transportation planning. While the report focuses on systems planning, it provides an outline of methods used as well as the results of surveys with State and regional practitioners.
    > http://nepa.fhwa.dot.gov/ReNEPA/ReNepa.nsf
- Barberio, Gina, Rachael Barolsky, Michael Culp, Robert Ritter, "PEL – A Path to Streamlining and Stewardship," March, 2008.
    > An article describing the FHWA's Planning and Environment Linkages program with particular emphasis on corridor and systems-level activities. Examples used include Southeast Michigan Council of Governments, North Carolina and Missouri DOTs,

among others.
http://www.fhwa.dot.gov/publications/publicroads/08mar/01.cfm

- Culp, Michael, and Rob Ritter, "Planning and Environment Linkages: Overview and Examples," January 13, 2008.
    A presentation from the Transportation Research Board workshop on environmental analysis that describes PEL while using examples from Riverside County, CA, Colorado, North Carolina, and North Central Texas Council of Governments.
    http://www.environment.fhwa.dot.gov/integ/resources_publications_MCRR.asp

- U.S. Department of Transportation, "Environmental Mitigation in Transportation Planning: Case Studies in Meeting SAFETEA-LU Section 6001 Requirements," October 2009.
    This report presents findings from nine case studies examining various environmental mitigation activities that transportation agencies have taken to respond to new requirements under SAFETEA-LU.
    http://www.environment.fhwa.dot.gov/integ/pubcase_6001.asp

- Council on Environmental Quality, "NEPA Task Force Report," 2002.
    This report provides recommendations for how to better integrate NEPA into Federal agency decisionmaking and to making the NEPA process more effective, efficient and timely.
    http://ceq.hss.doe.gov/ntf/index.html

- U.S. Environmental Protection Agency, "National Environmental Policy Act," Web site.
    This Web site is a resource for information about NEPA, compliance, and processes for submitting Environmental Impact Statements (EISs).
    http://www.epa.gov/compliance/nepa/index.html

- Federal Transit Administration – Environmental Analysis and Review Web site
    The Environmental Analysis and Review Web site provides links to information about the NEPA process, environmental mitigation, professional development and other resources.
    http://fta.dot.gov/planning/planning_environment_5222.html

C.4 STATE PROJECT EXAMPLES

- Montana Department of Transportation, "Libby North Corridor Study Public Information Site."
    This Web site provides access to the Libby North Corridor Study documents as well as information on public involvement and project status updates.
    http://www.mdt.mt.gov/pubinvolve/libby/

- Colorado Department of Transportation, "Parker Road Corridor Study," February 2009.
    Corridor plan Web site with information on the project and links to all final reports.
    http://www.parkeronline.org/index.aspx?NID=571

- North Central Texas Council of Governments "Streamlined Project Delivery – Regional Outer Loop."
    Project Web site for the Regional Outer Loop corridor study providing information on outreach meetings and preliminary documents.
    http://www.nctcog.org/trans/spd/roadway/OLoop/index.asp

- Idaho Department of Transportation, "Ashton to Montana State Line US-20 Corridor Plan," July, 2006.
    The final plan document for the US-20 corridor.
    http://itd.idaho.gov/planning/corridor/atmsl/FinalJuly.pdf
- Pennsylvania Department of Transportation "I-83 Master Plan Web site,"
    The project Web site with planning documents, contact information, and public outreach.
    http://www.i-83beltway.com/mp/index.html
- Delaware Valley Regional Planning Commission (DVRPC) Corridor Planning
    The corridor planning Web site with planning documents, contact information, and corridor planning vision for the DVRPC.
    http://www.dvrpc.org/corridors/
- City of San Diego, California Rosencrans Corridor Mobility Study
    The final plan document for the Rosencrans Corridor.
    http://www.dot.ca.gov/dist11/departments/planning/pdfs/systplan/RosecransCorridorStudyNoAppendicesCBTPGrantFebruary2010.pdf
- Virginia DOT and Thomas Jefferson Planning District Commission (TJPDC) Places29
    The plan Web site for Places29, which incorporates placemaking and transportation solutions.
    www.tjpdc.org/transportation/places_29.asp
- Alaska DOT&PF Auke Bay Corridor Study
    The project Web site for the Auke Bay Corridor Study.
    http://dot.alaska.gov/stwdplng/projectinfo/ser/abcorr/index.shtml
- WSDOT Interstate-405 Corridor Study
    The project Web site for the Interstate-405 Corridor Study.
    http://www.wsdot.wa.gov/projects/i405/
- Arkansas' Ecoregion-Based Approach to Wetlands Mitigation
    This Web site provides a case study and contact information.
    http://environment.fhwa.dot.gov/ecosystems/eei/ar05.asp
- Arizona DOT North-South Corridor Study
    The project Web site for the North-South Corridor Study.
    http://www.azdot.gov/Highways/Projects/NorthSouthCorridorStudy/Index.asp
- Maryland Intercounty Connector Project NOI
    Federal Register / Vol. 68, No. 106 /Tuesday, June 3, 2003.
    http://edocket.access.gpo.gov/2003/pdf/03-13794.pdf
- Washington, DC South Capitol Street Roadway Improvement and Bridge Replacement Project NOI
    Federal Register / Vol. 70, No. 79 / Tuesday, April 26, 2005.
    http://edocket.access.gpo.gov/2005/pdf/05-8330.pdf
- Connecticut I-95 Improvement Project NOI
    Federal Register / Vol. 72, No. 162 / Wednesday, August 22, 2007.
    http://edocket.access.gpo.gov/2007/pdf/07-4109.pdf

C.5 Related Research

- PB Americas, Inc. and Perkins Coie LLP. "Guidelines on the Use of Tiered Environmental Impact Statements for Transportation Projects." June 2009.
    > Prepared as part of NCHRP Project 25-25 (Task 38), the report provides guidelines on the use of tiering in the NEPA process. It is based on a review of over 20 tiered NEPA studies done between 1999 and 2008 for highway, transit, and passenger rail projects.
    > http://144.171.11.40/cmsfeed/TRBNetProjectDisplay.asp?ProjectID=1656

- Geiselbrecht, Tina Collier, Michelle Meaux, "Linking the National Environmental Policy Act and Planning: Case Study in Central Texas," 2008.
    > This case study examines the Capital Area Metropolitan Planning Organization (CAMPO) in Texas and its outreach to agencies not typically included in long range planning. The collaboration can be used to help with the NEPA environmental review process.
    > http://trb.metapress.com/content/d8m6366011718m55

- American Association of State Highway and Transportation Officials, "Using the Transportation Planning Process to Support the NEPA Process," 2008.
    > This handbook helps practitioners with linking transportation planning and NEPA processes and includes specifically a section on defining corridor-level goals.
    > http://environment.transportation.org/pdf/programs/practitioners_handbook10.pdf

- Galligan, Donald C., "Corridor Planning and the Integration of NEPA," Transportation Research Board, 2002.
    > This paper evaluates corridor planning as a tool for DOTs and MPOs, and also examines case study corridors and their use of NEPA in their planning process.
    > http://pubsindex.trb.org/view.aspx?id=804230

www.ingramcontent.com/pod-product-compliance
Lightning Source LLC
Chambersburg PA
CBHW081858170526
45167CB00007B/3061